Enumerations

Data and Literary Study

ANDREW PIPER

The University of Chicago Press   Chicago and London

The University of Chicago Press, Chicago 60637
The University of Chicago Press, Ltd., London
© 2018 by The University of Chicago.
All rights reserved. No part of this book may be used or reproduced
in any manner whatsoever without written permission, except in
the case of brief quotations in critical articles and reviews. For more
information, contact the University of Chicago Press, 1427 East 60th Street,
Chicago, IL 60637.
Published 2018
Printed in the United States of America

27 26 25 24 23 22 21 20 19 18   1 2 3 4 5

ISBN-13: 978-0-226-56861-4 (cloth)
ISBN-13: 978-0-226-56875-1 (paper)
ISBN-13: 978-0-226-56889-8 (e-book)
DOI: https://doi.org/10.7208/chicago/9780226568898.001.0001

The University of Chicago Press gratefully acknowledges the
generous support of the Social Sciences and Humanities Research Council
of Canada toward the publication of this book.

Library of Congress Cataloging-in-Publication Data
Names: Piper, Andrew, 1973– author.
Title: Enumerations : data and literary study / Andrew Piper.
Description: Chicago ; London : The University of Chicago Press, 2018. |
    Includes bibliographical references and index.
Identifiers: LCCN 2018003977 | ISBN 9780226568614 (cloth : alk. paper) |
    ISBN 9780226568751 (pbk. : alk. paper) | ISBN 9780226568898 (e-book)
Subjects: LCSH: Criticism—Statistical methods. | Criticism—Data processing. |
    Digital humanities.
Classification: LCC PN98.E4 P56 2018 | DDC 801/.95—dc23
LC record available at https://lccn.loc.gov/2018003977

♾ This paper meets the requirements of ANSI/NISO Z39.48–1992
(Permanence of Paper).

to my parents

Contents

Preface ix

Introduction (Reading's Refrain) 1
1   Punctuation (Opposition) 22
2   Plot (Lack) 42
3   Topoi (Dispersion) 66
4   Fictionality (Sense) 94
5   Characterization (Constraint) 118
6   Corpus (Vulnerability) 147
    Conclusion (Implications) 178

Acknowledgments 187
Appendix A 191
Appendix B 199
Data Sets 203
Notes 205
Index 241

Preface

This book is part of a longer exploration of the relationship between technology and reading, one that has occupied me for most of my career. In *Dreaming in Books*, I studied how romantic literature made sense of the bibliographic reorganizations that were sweeping across Europe and North America at the turn of the nineteenth century. Romanticism was, in this reading, a movement deeply invested in understanding material and technological changes that we are in many ways still grappling with. *Book Was There* sought to understand the recent technological upheavals around books by paying attention to more embodied dimensions of reading. Whether it is the touch or sight of the page or the places and practices of note-taking, game-playing, sharing, storing, or consuming books, I wanted to show how these experiences differ profoundly between print and digital media. Finally, *Interacting with Print* turned to the ways historical actors engaged with their reading material to produce new kinds of social communities, new models of creativity, and new structures of knowledge. Written with twenty-two coauthors, *Interacting with Print* put theory into practice in an elaborate process of scholarly interactivity of its own.

This then is the intellectual background to the book you are reading. *Enumerations* is about how computation participates in the construction of meaning when we read. It argues that data and computation unquestionably have a role to play in understanding literature, but that the way we have so far approached this problem rests on a number of flawed premises. The notions of distance, bigness, or objectivity that are largely in circulation right now rely on overly binary models of reading, largely untethered from past practices. *Enumerations* tries to show how these frameworks do not adequately capture the nature of computational modeling and its place within the rich history of reading. We still do not have a clear picture of how emerging quantitative methods speak to the questions that matter within the discipline of literary studies. This book is an attempt to align new kinds of models with old kinds of questions.

As I began to think about why I was interested in the question of literary quantity, I realized that it marked an even more general continuum with previous concerns. It belonged to my abiding interest in understanding the commensurability of seemingly incommensurable things. Instead of exploring the relationship between words and their objects, or bodies and reading, or paper and electronic books, as I had done previously, here I have moved to the relationship between letter, number, and image (in the form of the diagram). However disparate, behind each of these efforts lies the idea of translation, the act of moving between languages, cultures, and mentalities, as a core practice, but also an ideal, for humanistic scholarship. The first book I ever published was a translation, and it occurs to me now that I have been writing under this sign ever since. In the back of my mind, I keep trying to imagine an alternative future where students are not dutifully apportioned into silos of numeracy and literacy, but are placed in a setting where these worldviews mix more fluidly and interchangeably.

As much as this book represents a continuum, it also marks a breaking point, both from my past work, but also in the sense of something being broken. The research for this book began concretely when I started "retraining" myself in the field of computational text analysis several years ago, combining the practice of computer programming with that of quantitative reasoning. As hard as this process was and continues to be, computation allowed me to gain two fundamental insights about our discipline that had so far been overlooked. The first is the pervasive quality of textual repetition. The vast bulk of any single text consists of elements that repeat themselves with great frequency. These repetitions in turn multiply out in the world, giving coherence to entire domains of writing, such as genres, periods, modes, topoi, and careers. And yet, we have had no way of accounting for this fact of recurrence. It was as though we had elected to orient ourselves around rare events to protect ourselves from the vast majority of textual features (not to mention texts themselves). Focusing on a single dash in Heinrich von Kleist makes sense only if you pretend that there are not tens of thousands of other ones floating around.

The second problem is what I discuss in the introduction as a science of generalization. Until recently, we have had no way of testing our insights across a broader collection of texts, to move from observations about individual novels to arguments about things like *the novel*. And yet, we make these generalizations all the time. Indeed, one could argue that generalization is a crucial aspect to any scholarly method. It is what allows us to identify the significance of a particular instance as well as the social and historical significance of some larger set of practices. It is how we move between part and whole.

As recent research has begun to suggest, those wholes have been expanding for some time. The scale of our categories (world literature, new media, post-canon) has been matched by an increasing attention to social critique, to questions of worldly "mattering." And yet, our methods have remained largely unchanged. I will never forget the moment when I realized that the usual answer our field offers to initiates when faced with this problem—read more!—suddenly seemed incredibly, even senselessly, insufficient. As the Enlightenment scholar Johann Hamann once said, the imperative to read more feels like the punishment of carrying water through a sieve meted out to the daughters of Danaus. More reading could never by itself provide the evidentiary foundations to make categorical arguments—whether about Romanticism, modernity, the book, the novel, or even literature. We require some way of traversing scales, of testing our individual insights and observations against a set of texts that is more representative of the category about which we are speaking, especially as our analytical scales keep expanding. It is clear that *more* needs to be replaced by something we might call *method*. I had seen the crack in the table.

I take this expression from a short 16 mm film produced by the artist Paul Sietsema ("Anticultural Positions" [2010]). In it, we see a number of still photographs of the surfaces upon which he worked while making the photographs and paintings used in an earlier film ("Figure 3" [2008]), one that was itself largely concerned with the representation of surfaces, like paper and pottery. What makes these films so moving is Sietsema's attention to the fissures and lines that corrugate any surface when looked at closely, the way he sees the furrows of surface. At one point, he shows us a marble tabletop in his studio, in which we see a slight crack. Behind the pristine surfaces of knowledge, the foundations upon which something else is made, Sietsema reminds us that we also need to see the cracks, the places of vulnerability within the whole. It was these cracks in the otherwise smooth surface of reading that computation had allowed me to see.

Sietsema's imagery offers a useful metaphor in another sense, too, because it draws attention to the visibility of the materials we use when we create. Filming the surfaces upon which the objects of a film are made highlights the infrastructures of how we know things. As I argue in the introduction, one of the affordances of computational reading is the way it makes the critical project more legible than has traditionally been the case. While there will always be tacit dimensions to knowledge (as Michael Polanyi was the first to remind us), computation can be far more exo- than endoskeletal when compared with inherited critical practices. It is in this spirit that I have tried to make as much of the data and code used in this book publicly available. This includes over 7,000 lines of code (paltry for some, elephantine for me), as well as hundreds of tables of derived data from the primary data sets, all organized by chapter. While many of the primary data sets, which are described more fully in the appendix, cannot be shared, due to copyright, I provide code and tables of metadata about the collections for you to extract and build your own versions of them (or at least understand what has been included in them). I am trying to set a standard of reproducibility that will, I hope, gradually become more of a norm.

Throughout, I have adopted the convention of describing each model or calculation referenced in the text in the notes. I generally favor plain-language descriptions of models over formulas and equations. The notes also contain a subsequent reference to the accompanying piece of code beyond the book (i.e., see script 1.1), where the full implementation of a model can be reviewed in greater detail. In doing so, I am trying to strike a balance between the conventions of the humanities, which emphasize reading as a form of knowledge in its own right, and those of more quantitative disciplines, which put all the formulas and tables up front. Others may want a different approach, but my hope is that this allows for a thoughtful reading experience as well as the ability to replicate a model. It maintains the spirit of the foot- or endnote as a paratextual space with a difference. You are free to use the code for your own purposes or to try to reproduce the results I put forth here. I make no claims to elegance in programming, but I am confident that the scripts *work*, at least as of today. Durability has taken on a whole new scale of meaning when seen against the long timescales of bibliographic preservation.

At its heart, this book is an attempt to bridge two very different intellectual worlds and ways of thinking and reading. It would not have been possible without much generosity on the part of people from both of these worlds, some of whom I explicitly name in the acknowledgments or notes, but there are many more. The field is too diverse to be captured by a single proper name. We would do well to acknowledge that. Throughout the research and writing of this book, I have received tremendous amounts of help from others. This work is unquestionably more collective than traditional scholarship in the humanities. But it is also more bootstrapped, to use a computing term borrowed from the world of horses, in the sense of being more improvisational. Much of what I have learned has been acquired through the meandering and chance encounters of someone making his way through new terrain. As Adam Hammond has argued, there is a DIY quality to programming and computational criticism that is inspiring and pedagogically encouraging. This book wants to convince you that if you are not already doing so, then you too can enter into the world of computational reading.

If we are going to foster this sense of exploration (and the potential for getting lost), then we ultimately need a more flexible model of what it means to be an expert. Alongside the expertise of specialization, we need to value the expertise of synthesis and mediation, what it means to speak two different languages, or codes, or embody two different mentalities simultaneously. This book is dedicated to all of those people who don't feel at home inside something, whether it is a culture, a club, or a discipline, and instead who think there is something important to be discovered, something novel and consequential, in the spaces between.

Introduction
(Reading's Refrain)

"What is the sum of the text?" ROLAND BARTHES

Repetitions

In *The Jew's Daughter*, a now classic work of electronic fiction by Judd Morrissey, we see an image of a single, static page. As the cursor passes over a highlighted word, portions of the page change, while some of the words stay the same. Unlike the turning of the page in a book, where a visual space is entirely overwritten, here only parts of the page change, even as it maintains its overall formal stability. The pages of *The Jew's Daughter*—if we can call them that—not only follow one another in a linear sequence. They are also woven into one another. We might say, drawing on a bibliographic metaphor, that they are interleaved.

The title of Morrissey's work was taken from the name of a ballad sung in James Joyce's *Ulysses*. In one sense, it performs a familiar literary gesture where a more canonical work is cited by a later one, much as Joyce's own work had done. At the same time, in the invocation of the genre of the ballad, *The Jew's Daughter* also highlights the poetic device of the refrain as a cornerstone of literary expression, if not of the experience of reading itself. In this, it is perhaps drawing on the account from Augustine's *Confessions* in which the child's nearby refrain initiates the author's personal conversion, one that takes the form of the repetitive command "Take it and read, take it and read." Rhetoric and medium, the page and the refrain, recapitulate one another through their mutual investment in repetition and transformation. They too are interleaved.

In taking seriously the long history of reading's technics—in foregrounding this elementary association between reading and the refrain—Morrissey's work invites us to think about reading in a profoundly new way. It asks us to reflect on the meaning of reading as a form of rereading, not in the Augustinian sense as that which comes after reading, as a form of personal transformation. Rather, Morrissey's work asks us to think about the meaning of rereading *within* reading, to understand the redundancy that is reading. When an average of 56% of words repeat themselves with every turn of a novel's page, what is the meaning, Morrissey's work is asking, of the quantities that underlie such repetitions?[1] What does it mean to read the same thing again and again?

Ever since Pliny the Younger said, "Read much, not many," debates about quantity have been central to debates about reading and literature. To read the *right amount* has often carried with it a moral imperative.[2] At the same time, the sheer volume of reading material available at any given time has often been seen as culturally significant in its own right. A great deal of research in the fields of book history and bibliography has drawn attention over the years to how the quantity of physical documents has shaped the literary and intellectual landscape.[3] According to this point of view, more things matter.

This book offers a new perspective on the significance of quantity for the study of literature. Inspired by the emerging fields of natural language processing, machine learning, and text and data mining, as well as a host of colleagues beginning to work in this area, *Enumerations* explores the quantitative dimensions within texts, the ways in which the repetitions of language lend meaning to our experience as readers. As Gilles Deleuze once remarked, the literary critic has traditionally been a purveyor of rarity, watching over the singular achievements of singular individuals, much like the gatekeeper of Kafka's parable "Before the Law."[4] And yet so much of the activity of reading is built upon repetition, all of the ways that words, and the ideas and feelings they give birth to, have quantitative dimensions. We are now able to move beyond simply counting books to measuring the complex features that reside within and between them. But these developments raise a fundamental, and as yet unanswered, question: What is the meaning of literary quantity?

Enumerations is an attempt to answer that "meaning question," as Alan Liu has called it, across some of the most elementary dimensions of literature.[5] In keeping with the embryonic state of the field, it concentrates on the building blocks of literary study, from the role of punctuation in poetry, to the matter of emplotment in novels, to the dispersion of topics, the behavior of characters, to the nature of fictional language and the shape of a poet's career. It does so by assembling new kinds of data that represent different literary forms over the past two centuries. This includes over 230,000 poems, 15,000 novels, and another 12,000 works of nonfiction.[6] Through data, it attempts to tell the deep story of these elementary literary features, where by "deep story" I mean all of the ways that cultural practices manifest themselves in repetitive, often predictable, and sometimes excessive ways.[7] Paying attention to quantity reveals the grooves and channels of cultural expression, the deep connections among words, ideas, and forms.[8] It brings us back in many ways to an originary sense of culture as an agricultural form of cultivation. Repetitions etch themselves into the social fabric, like so many furrows in the ground. They are often invisible, because so common. It is only when we take into account these repetitions—in my case 3,388,230 punctuation marks of twentieth-century poetry, 1.4 billion words of novels over the past two centuries, or 650,000 fictional characters that populate the nineteenth century—that we are able to see the deep story of poetry, the novel, fictionality, character, the commonplace, and the writer's career begin to emerge.

In drawing attention to the quantities of literature, *Enumerations* aims first and foremost to rethink how the computational study of literature has initially been framed.[9] The seemingly endless array of debates surrounding the field—indeed, the way "debates" has emerged as a kind of default genre within the field—has led to hardened and quite often deeply anachronistic notions of disciplinary history. We are talking not only past each other, but also past the past itself, overlooking the numerous ways that this mode of inquiry grows out of a variety of different traditions.[10] The emphasis on novelty, but also bigness, empiricism and, as we will see, overly simplified and often binary models of reading, has led us to miss the important ways that computational reading is inevitably tied to the norms and practices of the past. In misapprehending this disciplinary inheritance, we too easily misjudge the ways in which it does indeed offer distinctive challenges to disciplinary norms in the present.

The notion of repetition behind Morrissey's digital page is a case in point. Ever since Augustine, the act of rereading has served as a deeply effective practice of cultural stabilization.[11] It has stood alongside a variety of other repetitive cultural practices, such as copying, reprinting, note-taking, commonplacing, anthologizing, canonizing, and memorizing, which have all assumed cultural importance and have been the focus of much academic research.¹² This book argues that the repetitions of language are no less important, even if until now largely overlooked. Quantity signals, but it also distinguishes and maintains. If there is a basic dogma against which this book argues, it is that literature is not founded on the rare and the singular, but rather the common and the collective, the fabric of repetition from which it is made.¹³ In this, it situates computational research within a much longer tradition of literary study that attends to questions of cultural distinction and durability, but does so in new ways.¹⁴

For many in our field, the introduction of quantity into the study of literature has been seen as nothing short of an act of intellectual colonization. Numeracy's rise signals literacy's eclipse, and with it a host of highly charged concepts like subjectivity, individuality, creativity, or even agency. Language plays subjective foil to number's objectivity. According to a critic like Friedrich Kittler, there has been an ongoing 2,000-year battle between the forces of literacy and numeracy.¹⁵ The impoverished view that this takes on the history of literature and thought should be obvious. The list of philosophers who were mathematically trained is long, from Leibniz and Descartes to Ludwig Wittgenstein, just as the centrality of number to literature is equally rich. That there are nine circles of hell in Dante's *Inferno*, 100 stories in Boccaccio's *Decameron*, 365 chapters in Hugo's *Les Miserables*, 108 lines in Poe's *The Raven*, and a genre of poetry with exactly 14 lines that has lasted for over half a millennium (not to mention the entire field of prosody) indicates just some of the ways that quantity has been an essential component of literary meaning since its inception.

Far from seeing the computational turn in literary studies as something distinctively new or even alien, we can and should understand it as part of the history of humanism itself. Humanism was in many ways founded upon the notion of studying linguistic and material differences, a practice of fostering the ability to understand the ways in which texts differ between times and places. Translation would emerge as one of its core practices as well as ideals. The knowledge gained in moving between languages, historical epochs, and systems of writing was seen as the highest form of knowledge. Crossing the divide of textual and linguistic difference was a means of potentially crossing the divide to something more spiritually transcendent. Erasmus's bilingual New Testament might be considered to be one of this tradition's most important founding documents.

Today, there is a new translational imperative at work, one that aims to move between letters and numbers. Translating texts into quantities has emerged as the overwhelming feature of our cultural moment. Rather than see this as a kind of fallen state, I think we would do well to reposition it within a longer tradition of translational humanism, to see it as part of an ongoing intellectual drama that tries to understand the act of commensuration, of making different sign systems compatible with one another. Seen in this way, the literate and the numerate are not agons engaged in a duel. They are two integral components of a more holistic understanding of human mentality. In its most general sense, this book is an attempt to think through the value of translating between letters and numbers as a newly vital form of humanistic thought.

If our debates have missed some of the more important ways that computational criticism is imbedded in long-standing disciplinary practices, it has also obscured the need for more interdisciplinary conversation. Most of the methods for the computational study of literature are emerging from other disciplines, indeed from other faculties altogether. Our conversations have to date been far too hermetic. Where the humanistic side of digital humanities has been too removed from the norms and practices of the past, the digital side has been too detached from the present of computational research. Different researchers have emphasized different intellectual traditions from which to draw on. For James English, Ted Underwood, Hoyt Long, and Richard Jean So, sociology and the social sciences offer an important framework through which to understand the computational turn within literary studies.[16] For researchers like Berenike Herrmann, Gerhard Lauer, and Winfried Menninghaus associated with the International Society for the Empirical Study of Literature, it is the cognitive sciences and their emphasis on experimental method that provide a generative new framework through which to understand literature's social and aesthetic importance.[17] As you will see, this book is far more indebted to methods emerging from fields like computational linguistics and information science, with their focus on the algorithmic and statistical understanding of language.[18] What is important in all cases is a deeper immersion in the literature and methods that these fields have to offer.

The reason to do so, however, is not to adopt uncritically another field's wares (or ultimately be subsumed by them), but rather to better understand the *limitations* of transferring one discipline's methodological apparatus onto another, and in the process improve *both*. More collaboration and cross-talk is needed to better understand not only how these models might be able to address the kinds of questions we care about, but also the problematic assumptions we see in their current applications out in the world. Concepts like machine learning, artificial intelligence, and data science will undoubtedly change for the better as more humanists enter into the conversation.

Enumerations is the product of many years of such conversations. It is born from listening, but also the work of transferring, and the inspirations and frustrations that accompany inhabiting this space in between. Ultimately it is about moving us away from a polemical mind-set to an analytical one, to continue the work that is already underway of presenting concrete insights into how the study of literary quantity changes our understanding of literature. Each of the chapters attempts to intervene in ongoing debates in literary study, whether it be twentieth-century poetics, the history of the novel, the study of character, or theories of authorship. And they do so through the application of a variety of techniques that will gradually grow in sophistication as the book progresses. The literary elements that the book studies are thus mirrored in the methodological elements used to understand them. As I move from extracting regular expressions ("grep") like punctuation in chapter one, to using vector space models and social networks to approximate plot in chapter two, to examining topic models in chapter three, to using machine learning in chapter four, to dependency parsing and co-reference resolution to study characters in chapter five, to the use of more nested models that measure the extent of a poet's "vulnerability" in the final chapter, this book is aimed at bringing readers into this world, but always with an eye to concerns within literary scholarship. It is less a handbook and more an extended demonstration of the ways in which computation, when applied critically and creatively, can confirm, revise, but also invent new narratives about literary history. Rather than continue to make arguments for and against computational criticism, we need to let the arguments of our research make the case for why this work is important. But before we get there, we first need a sense of the conceptual transformations that this work entails.

Implications

Writing in exile in the 1940s, the German Romance philologist Erich Auerbach produced one of the landmark works of twentieth-century literary criticism. *Mimesis: The Representation of Reality in Western Literature* was monumental in both its scope—it covered texts from Homer to Virginia Woolf—and its categorical claims. It had as its focus "mimesis," "reality," and all of "Western Literature." It served as a potent warning to all aspiring philologists, like the terrible inscription that hung over the gates of hell in Dante's *Inferno*, not incidentally one of the books in Auerbach's massive oeuvre: "Abandon all hope, ye who enter here." Who could ever claim to possess such erudition?

Much recent work has been done to debunk Auerbach's self-portrait as an isolated exile.[19] Istanbul was hardly the cultural backwater that he made it out to be (Auerbach had after all last taught in Marburg, population ca. 30,000). But no one, to my knowledge, has ever tried to debunk Auerbach's erudition, his immense learnedness. Who would ever presume to have read more than Auerbach? But what if he had not read enough?

Surprising as it may sound, Auerbach's work offers a cautionary tale for the future of literary studies, but not because of its feigned orientalism, the way it establishes a too-convenient binary between East and West. Rather, its importance lies in the way it more universally inscribes exile as a foundational condition of literary criticism. *Mimesis* dramatized, perhaps more than any other work in the field before or since, the metonymical crisis that lay at the core of literary criticism, an incommensurable relationship between part and whole. Every chapter begins with a passage from a "great work," where the singular instance of text is meant to stand for the singular nature of the work under review, which in turn stands for all of "occidental literature." Behind all of this lay reality itself. The literary critic, stationed at a distance, captured the inaccessibility of the cultural whole to which his knowledge aspired. "You grasp the spirit that you resemble," says the Earth Spirit to Faust just prior to disappearing, as the whole slips once again through his fingers. This from one of the great works of literature about the limitations of humanist knowledge. It is telling that Auerbach did not include it in his treatise.

Ever since Auerbach, we have been replaying this epistemological tragedy in the field of literary criticism and the humanities more generally. Whether it is the anthropologist's notebook, the historian's archive, the media theoretician's screen, the art historian's collection, or the literary historian's book, each of these media is in its own way a flawed portal to understanding something larger than itself. They each fail, in a word, to *generalize*. One cannot make visible the linkages between the particularities of a given document and the larger context to which it belongs with anything close to the rigor and specification applied to the individual document itself. Like Auerbach in Istanbul, the detailed attention to the particulars of language exiles us from an understanding of the representativeness of our own evidence.

Despite its title, *Mimesis* was importantly not about the "representation" of reality, but what Auerbach called "represented reality" [*dargestellte Wirklichkeit*]. The translation leads us slightly astray. There were not two separate categories (representation and reality) for Auerbach, but a single unified one. Reality was always mediated through some kind of construction, one that was historically varying in nature. This was one of the profound insights of Auerbach's work. And yet even *Mimesis*'s insistence on the representedness of reality was still blind to the *representativeness* of its own examples. Whether it was the passages that stood for the works from which they were excerpted or the works that stood for the culture from which they were drawn, there was no way for Auerbach to address the fraught relationship between part and whole. How well did these local examples account for the great works from which they came, or the even more general phenomenon of Occidental literature? How would it be possible to generalize under such terms?

Some might respond that we should simply abandon the act of generalization and go back to what we do best: studying the particulars of cultural practice. "The task of understanding then," write Stephen Greenblatt and Catherine Gallagher in their manual *Practicing New Historicism*, "depends not on the extraction of an abstract set of principles, and still less on the application of a theoretical model, but rather an encounter with the singular, the specific, and the individual."[20] Literary criticism, so this version goes, is at its best when it is attuned to the particulars, the profound singularity, of any cultural document or practice. We are freed from the impossible burden of reading everything because all that matters is the power of the local insight. Who has not written in the margin of many a student paper, "too general"? The very thing we seem to value most in our teaching and research is an attention to specificity.

And yet in one of the great paradoxes of intellectual history, it was precisely New Historicism that drew attention to a document's "social energy," the complex ways in which it was imbedded in a variety of social practices, all the while defending the value of the anecdotal nature of evidence ("the method of the Luminous Detail," in Ezra Pound's words). Decades worth of critical theory have shown us the value of thinking more generally about culture and the extent to which recourse to "particularity" serves as a convenient fiction, one with very clear political overtones. It is the meme of the heroic individual, like Auerbach in exile or Shakespeare at the Globe, freed from the constraints of context. But how could this ever be possible? What are the interests behind the invocation of such difference? Why do we forget our implicatedness at the very moment it matters most?

This then is one of the major contributions, and challenges, of computational literary study. Rather than abandon generalization, it sets as its task a reflection on the representativeness of its own evidence. It aims to apply Auerbach's notion of "represented reality" to our critical practices and develop methods that are as rigorous and specific in their assessment of some larger whole as they are of some local engagement. While we spend a great deal of time training readers to be more attentive to what's in a text ("too general"), we spend precious little time reflecting on the process of generalization itself, of how we move from the luminous detail to arguments about the larger social contexts in which those details are imbedded. We lack a science of generalization.

One of the core concepts that can assist us in this process and that has so far been missing from recent debates surrounding the digital humanities is that of *modeling*. As we will see, the notion of modeling is explicit within a variety of the approaches used here (vector space models, topic models, predictive models). There is a great deal of literature on the role of modeling in knowledge creation, and it should become required reading for our field.[21] As historians and philosophers have pointed out, models are first and foremost representations, miniatures that mediate between ourselves and our observations. They do not reproduce the book or literary field as it is, but rely on a logic of what the early twentieth-century philosopher Hans Vaihinger called the "as if [*als ob*]."[22] For Vaihinger and his intellectual descendants, fictionality was a core part of epistemology.[23]

Thinking about modeling realigns our focus around the ways in which knowledge is always mediated, the small details through which our insights about large numbers of texts are constructed. Much of the early discourse surrounding the computational analysis of literature has inevitably focused on notions of distance or bigness, on a vocabulary of transcendence or the macrocosm. Similar to the origins of computing culture or literary studies itself, the initial emphasis has been to prioritize a sense of communion with something greater than ourselves.[24] Focusing on models, thinking small in order to think big, moves us away from this binary logic of size—where bigger is always better—toward one of representation.[25] As we begin to reflect on the *representativeness* of our evidence, large and small, close and distant, become interwoven (fig. 0.1). There is a far more circular nature to computational reading than has so far been acknowledged. Where the social sciences often speak in terms of "samples" and "bias," the notion of "representativeness" suggests that there is not ultimately some stable, knowable whole against which one must limit one's "bias." Instead, it foregrounds the ways that the ideas, meanings, and feelings of literature are never capturable once and for all, but are themselves always forms of Auerbachian represented reality [*dargestellte Wirklichkeit*]. Models foreground the constructedness of knowledge and the observer's place within it.

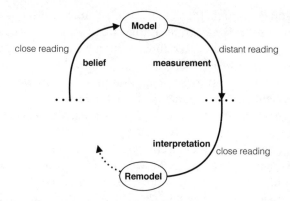

0.1   Schematic representation of the practice of literary modeling.

The data sets that I use here are thus not to be confused with some stable macrocosm, a larger whole to which they unproblematically gesture. They too are constructions, just as much as Auerbach's collection was. When I talk about twentieth-century poetry, the novel of lack, the topos of life and death, the introverted heroine of female novelists, the phenomenological investment of fiction, or the nature of late poetic style, in each case my observations are dependent on the collections that I have assembled for these purposes. Future work will want to explore the extent to which these and other observations depend on different kinds of collections. We need to better understand the interactions between methods and archives.[26]

If models help reinsert us within a more hermeneutic stance to the past, they also introduce an important notion of craft. They prioritize *explicitness* as a core value behind literary study though, as philosophers of science have pointed out, there are still always "tacit dimensions" to any craft.[27] Models manifest. They help undo the black box of critical charisma. The antipode to modeling is not subtlety or complexity, but personal authority. Models replace charisma as the guiding vehicle of generalization. They diffuse power away from the persona (the proper name) into a more dispersed array of technologies, techniques, and practices among which the individual is enmeshed. Modeling is a form of explication.

When I construct a model in this book, I reference how many documents are being considered, what my beliefs are in approaching those documents, and how I am implementing those beliefs. In one sense, it is not that dissimilar from the normal procedures of literary studies, where I gradually develop a thesis as I come into contact with more and more documents. What is different is the way a model allows me to negotiate between part and whole (where the whole is of course never *the* whole, but some representation of it). In traditional critical practices, we only ever hear about the passages and works that fit our thesis (the parts). We almost *never* hear about the ones that did not, how many were considered, and how prevalent the phenomenon is that we are observing more generally. In building a model, the literary critic undergoes the same circular process of developing a thesis—call it a hypothesis—but then tests this hypothesis against a larger set of documents (or multiple sets). I might have a belief about what might happen—this is after all why it is called a hypothesis—but I can never know in advance. Unlike the imperious pronouncements of the literary critic who is only ever right, there is an element of uncertainty, error, and chance built into the task of computational criticism.

There is understandably an aspect of disenchantment about all of this, as the computational critic lays bare as much as possible of his or her intellectual process. The magic of critical insight is dispelled in favor of the craftsmanship of model-building. But there is something more consensual and less agonistic about the practice of modeling as well. Algorithms require that we define every step of our intellectual process. In their simplest sense, they are nothing more than a set of instructions. Humanists have traditionally been opaque about their methods—about what evidence has been considered and what is considered as evidence. Computation, too, is no stranger to opacity.[28] But done well and done *openly*, mediating knowledge through computational modeling is neither an enclosure of knowledge nor an act of dehumanization. At its best, it can be understood as a practice of mirroring, one where we show ourselves the cognitive steps we are taking to arrive at an argument about literature, what Walter Benjamin might have called an awareness of a text's "lexical unconscious." Computation in this sense is not fast. Rather, it slows us down and forces us to be more self-reflective. But it also allows others to participate in the process. The tools and information for mediating disagreement have been made mutually available. Modeling makes the study of literature more architectonic and less agonistic, more social and collective. Unlike the lone critic in exile, modeling encodes a togetherness to reading. As Benjamin intuited about the work of art, there is a basic politics of reproducibility at work in computational criticism that has largely been overlooked in current debates.

Finally, models matter because of the way they implicate us within them. In this, they run strongly counter to much of the language of empiricism that has surrounded the initial rise of the field. We are always present in our models. As I try to show in the conclusion, we can devise ways of making ourselves and our practices even more explicit within the models we build. But models also push against dominant notions of "critique" or "resistance" that continue to inform much of traditional criticism as well. In both of these cases, whether the empirical or the critical, the observer and the material conditions of observation are dropped from view in favor of what is seen or found. Both are exilic in nature. Instead, models inscribe our beliefs within them. They do not say that something is true for all time, but that under these conditions I believe this to be true right now. Unlike the exiled reader of Auerbach's critical allegory, models are always situated, in both time and space. But they also do not reside entirely in that space of pure subjectivity, of the *I-believe*. They simultaneously construct statements about the world *as a world*. They have a testing, rather than resisting, relationship to the world. They question, but also affirm. There is a positivism to model-building that also needs to be recognized.

It is precisely this process of contingent world-making that I would argue lends the practice of computational modeling its critical force. As Gilles Deleuze writes of the diagram, so too of the literary model: "Every diagram is intersocial and constantly evolving. It never functions in order to represent a persisting world but produces a new kind of reality, a new model of truth. It is neither the subject of history, nor does it survey history. It *makes* history by unmaking preceding realities and significations, constituting hundreds of points of emergence or creativity, unexpected conjunctions or improbable continuums. It doubles history with a sense of continual evolution."[29] Like the diagram in Deleuze, computational models are intersocial in that they allow for others to participate in their making, but also constitute new relationships through their making. At the same time, they are not simply simulations of the world around them—mere representations—but constructions of new kinds of realities. They do not just represent history, but also *make* it. There is a constructive dimension to models that is fundamentally different from the outsider status of critique and its ethos of resistance or unmasking or the empirical project of producing some sort of stable, transcendent truth. Models are dynamic. In Bruno Latour's terms, they entangle us, moving us from a space of bibliographic intimacy or enlightenment to one of algorithmic implication.[30] They replace notions of the readerly revolution, the single radical change that was at the core of Augustinian rereading and that has arguably remained central to our discipline ever since, with one of resolution, a persistent encounter with differing levels of scale. Models are ongoing.

Distributions

In one of the great treatises on reading, *The Pleasure of the Text*, Roland Barthes recounts his discovery of the importance of what he calls the "*non-sentence.*" It is an insight he says he had one night while sitting in a bar. Listening to "all the languages within earshot," Barthes writes, "I myself was a public square, a *souk*; through me passed words, tiny syntagms, bits of formulae, and *no sentence formed*, as though that were the law of such language. This speech, at once very cultural and very savage, was above all lexical, sporadic; . . . it was: what is eternally, splendidly, *outside the sentence*. Then, potentially, all linguistics fell."[31] Once again we can see the orientalism of European criticism at work here, where the novel insight, the discovery of something other, is translated through the topos of the East (the Arab bazaar or *souk*, but also the "savage"). For Barthes, the sentence was a fundamentally hierarchical unit, full of syntactic dependencies, while the lexeme (or phoneme or morpheme) was primarily heterarchical. "The professor," Barthes reminds us, "is someone who finishes his sentences." According to Barthes, the sentence has traditionally served as the primary unit of literary analysis—professors do not just speak in sentences, but like to look at them—because the sentence is a mark of distinction. There is an authority encoded in the syntactical primacy of the text.

In thinking beyond the sentence, in thinking about a more distributed notion of language, Barthes was very much of his moment. A variety of theoretical positions would emerge during the period we now identify as post-structuralism that were deeply spatial in nature. Barthes's idea of structuration in *S/Z* or the rhizomatic or transversal reading of Deleuze and Guattari were two of the more notable examples, but so too was Kristeva's and Genette's thinking about intertextuality and the porous, reticulated boundaries that were imagined to exist between texts.

However counterintuitive it may seem, such notions are now deeply imbedded in the computational modeling of texts. The concept of "distributional semantics," which lies at the center of almost all approaches in the field of computational linguistics today, is in many ways a profound manifestation of post-structural textual theory. At base, the distributional hypothesis assumes four things: a) a word's meaning is tied to how often it occurs; b) a word's meaning is tied to how often it occurs with *other* words within a given context; c) these relationships are entirely contingent upon the scale of analysis; and d) these relationships can be rendered spatially to capture the semantic associations between them.[32] The distributional hypothesis is thus based on a cognitive assumption about the probabilistic way we assess meaning through language and a rhetorical assumption about the importance of spatial relationships for the construction of meaning.

In the first case, it assumes that we generate expectations about a word's meaning based on the frequency with which it occurs within a given linguistic context. Changes to these frequencies, as well as changes to their contexts, will change our assumptions about the word's or document's meaning.[33] In the language of cognitive psychologists, these changes will impact the "associative activation" that influences our judgments about language.[34] In the second case, it suggests that modeling the frequencies as coordinates in space, in what is known as a vector space model, reconstructs the word's or document's potential meaning, not as something fixed and definitive, but as positioned within a larger field of relations (called "term space" or "feature space").[35] There is a profound spatialization to reading that is implied through computational models that will be enacted throughout this book. In this, such methods draw on the long history of the "figure of speech" as a core way through which language has been thought to signify.

To better understand how this works in practice, we can build a vector space model of the relationships between the following three sentences taken from *The Sorrows of Young Werther*:

1. My dear friend what a thing is the heart of man!
2. I treat my poor heart like a sick child.
3. I have possessed that heart, that noble soul.

The first step is to transform each sentence into a distribution of word frequencies, giving us a total of twenty-two overall dimensions (table 0.1). Next, we can calculate the similarity of the sentences to each other across these twenty-two dimensions using a measure of spatial similarity (of which there are many). In this case, I will use what is known as cosine similarity, which measures the cosine of the angle between any two vectors.[36] Imagine plotting each sentence in twenty-two-dimensional space and drawing a line from some neutral starting point (zero) toward the sentence's coordinates. The angle between each sentence's vector is a representation of how semantically similar they are to each other (where a value of 1 means the documents are identical to each other and a value of 0 means they share no similarity at all). According to the model (table 0.2), the first and second sentences are the most similar because they

Table 0.1 Word frequencies of three sample sentences drawn from *The Sorrows of Young Werther*. Only a portion of the full table is shown.

	a	child	dear	friend	have	heart	i	is	like	man
Sentence 1	1	0	1	1	0	1	0	1	0	1
Sentence 2	1	1	0	0	0	1	1	0	1	0
Sentence 3	0	0	0	0	1	1	1	0	0	0

Table 0.2 Similarity matrix using cosine similarity for three sample sentences from *The Sorrows of Young Werther*

	Sentence 1	Sentence 2	Sentence 3
Sentence 1	1	0.302	0.095
Sentence 2	0.302	1	0.211
Sentence 3	0.095	0.211	1

share not only *heart* but also *my* and *a*. There is an emphasis on possession and generality in sentences one and two (*a* thing, *a* child), whereas the third sentence focuses more on agency (*I*) and specificity (*that* heart, *that* soul).

With just twenty-two dimensions, and only three sentences, this is obviously an overly simplistic model. And while I have relied on the use of words here, vector space models do not exclusively depend on them. As I show in chapter one, we can focus on punctuation as a feature of interest or, as in chapter six, we can mix features together, such as parts of speech, words, phonemes, and semantic relationships. One can also include more abstract features like topics or concepts, such as "lack" in chapter two, "life" in chapter three, or "introversion" in chapter five. Texts too can be broken up at any level of scale, from the sentence or phrase up to the individual work or an entire genre or period. There is a fundamental plurality to how texts can be modeled in a distributional framework, a lack of essential boundaries. But there is also a crucial aspect of multidimensionality, a way of seeing our objects of study as fluid configurations of multiple interacting components.

The standard criticism that is often leveled against computational models is that they take words "out of context," where context is assumed here to be the continuous flow of words in a sentence and the meaning that accrues from it. While it is true that distributional models take words out of *this* context (as in my table above), it is not the case that they remove context altogether. Rather, they recontextualize both words and documents within greater levels of scale. They implement Barthes's intuition about the importance of the space "outside the sentence." "We know that a text," writes Barthes, "is not a line of words releasing a single 'theological' meaning (the 'message' of the Author-God) but a multi-dimensional space in which a variety of writings, none of them original, blend and clash."[37] Computational models allow us to understand how an individual word (or feature) is being used relative to a larger context (the twentieth century, the nineteenth-century novel), just as they allow us to understand how an individual word (or feature) is being used in the context of other words and features. Part and whole are always kept simultaneously in view. To put it in more technical terms, vector space models translate the syntagmatic dimension of a word or passage's meaning—its local context—into a paradigmatic framework that is drawn from some larger distributional context.

It is worth pausing to look at an example of how the distributional method differs in practice from existing critical methods. When a critic argues, for example, that "*der Weg*, German for 'path,' 'road,' or 'way,' is a key word in [Kafka's *The Castle*]," we can see how we are in the realm of the symbolic rather than the distribution.[38] This one thing—the word/notion for *pathway* (and the way *word* and *notion* stand interchangeably for one another)—not only stands for the whole, but contains the whole within itself. It is the key that unlocks the novel's forbidden code. And yet, in a by now familiar twist, the access that is granted by this singular thing is then posited as fundamentally unknowable—a pathway, not a stable entity.[39] Like Kafka's castle, we as readers do not have full access to the meaning of what we are reading, nor even the methods through which such insight could potentially be derived. The value of the insight rests on a vacant authority, much like the architectural edifice at the heart of Kafka's text. Critical practice recapitulates its aesthetic object, as the critical reading preserves, or reassembles, the very power structures that Kafka's text attempts to lay bare.

A distributional approach would start from a very different premise. It would ask whether "Weg" is indeed excessively deployed in *The Castle* compared to other novels (it is, but only very weakly), and whether the semantic inflections of the word depart strongly from the expected norms of usage (they do not). How might we do this? In the first case, we compare the frequency of the word *Weg* with a range of expected values that depend on some larger collection, in this case a collection of 150 novels written in German and published over a century and a half (1770–1930).[40] If *Weg* appears far more (or less) often in Kafka than one would expect given the norms of usage within the novel more generally, then we can say that this word is significant for Kafka *with respect to this context*.

While simplistic in its implementation, the methodological reorientation at work here is not. Instead of saying this word matters (to me), it replaces the implied "to me" with reference to some larger social practice. This word matters *to the German novel* (so represented). This has profound implications for everything that follows, because we are no longer using our own judgments as benchmarks (the latent "to me"), but explicitly constructing the context through which something is seen as significant (and the means through which significance is assessed). Some degree of subjectivity has indeed been ceded in the process, but some has also been retained—in the generation of the question in the first place or the interpretation of significance in the second. It interweaves subjectivity with objects—materials, techniques, norms, and practices—that are outside ourselves, but that are in no way cut off from us. Subjectivity simply ceases being the benchmark of itself; it is now entangled with the quasi-objects surrounding it in the reading process.

In the second case, to understand whether the semantic inflections of the word in Kafka depart strongly from the expected norms of usage, we can build a vector space model, as discussed above.[41] Here, *Weg* is represented as a vector where the coordinates of that vector refer to the frequencies with which other words appear in proximity to it, in either individual novels or all novels in the collection. We can then compare these vectors to understand how semantically similar they are to each other. The more the frequencies differ—that is, the more the contexts of *Weg* differ for a given document relative to the whole—the more spatially dissimilar the semantic representations will appear to be from one another. Doing this for all of the novels in the collection across several thousand dimensions of terms, Kafka's novel looks neither strongly similar nor dissimilar to the general representation.[42]

The distributional approach thus allows us to understand significance in a relational way—with respect to a textual context (the words co-located with *Weg*), a social or historical context (German novels published in the long nineteenth century), and a particular method (cosine similarity). This is in many ways no different from the critic's approach, except in the latter case all of those contexts are latent (i.e., unspecified) and thus not open to view. At the same time, the distributional approach also allows us to *change* the context, in this case away from that of a single word and toward the document as a whole. It puts us in the resolutional stance I discussed above that is so central to the practice of modeling. Context is never fixed, but always perspectival. The distributional model moves us from a state of commitment or attachment, to one of contingent construction. Doing so allows us to assess other words, or word constellations, that are potentially more "significant" for Kafka's novel in terms of their excessive deployment (or abstention), the ways in which this novel can be located within its own distinctive semantic space when compared to the German novel more generally.

Such a perspective would lead us away from enigmatic keywords like *Weg* and toward more quotidian, yet arguably no less important, words like *aber* (but), *allerdings* (admittedly), *wahrscheinlich* (probably), *vielleicht* (perhaps), *offenbar* (obviously), and *freilich* (freely, in the sense of "of course" or "logically," as in "Dann hatte es aber freilich auch keinen Sinn" [then of course it also has no meaning]). All of these words are considerably more frequent in Kafka than in the German novel sample (occurring from two to as much as nine times more often).[43] It is arguably these words that were behind that assessment of *Weg*'s significance to the novel—the expanded context provides insights into the qualities that make Kafka's novel seem semantically unique within the longer tradition of the German-language novel.[44] While *Weg* is not used in a particularly unique way, it lives within a context that is deeply distinctive for the history of the novel. Together, these words capture that uniquely Kafkan universe of conditionality, the way everything is assumed to exist within a norm that can never fully be known, when something is "of course" or "obvious" or "probable," but no one ever knows why. They highlight the political and personal dangers of an unknowable linguistic context, the rise of expert systems that black-box their knowledge and create expectations to which not everyone has access.

Diagrams

Many of the early debates surrounding computational criticism have been shaped by a reliance on models of reading that are at once overly simplified and deeply binary in nature (distant/close, deep/shallow, critical/attached). In introducing the notions of repetitive, implicated, distributed, and finally diagrammatic reading, this book aims to reframe computational reading within a set of practices that are more attuned to the diversity that surrounds "reading" more generally. Paying attention to the repetitions that suffuse literary documents can make us aware of the ways that quantity impinges upon meaning, to think more translationally about traversing the mentalities that underlie literacy and numeracy. But it can also help move us away from historiographical models that rely too heavily on notions of breaks, ruptures, or singularities and toward questions of stability and duration, to see the deep grooves or furrows of literature, culture, or writers' lives. As I try to show in chapter three, methods like topic modeling can help us see long-standing semantic configurations that course through the novel's history, just as in chapter four I use machine learning to illustrate the remarkable stability surrounding the discourse of fiction for the past two centuries.

Focusing on the implicatedness of modeling, by turn, allows us to rethink our investments in either the nascent empiricism or residual subjectivity surrounding reading. It helps us see the intersections of these two positions, rather than their mutual exclusivity. Our recourse to imagining reading—and imaginative reading above all else—as a bulwark against the incursions of rationalization or as an incubator of pure subjectivity overlooks both the subjectivity inherent in modeling and a basic instrumentality to reading that has been operative for centuries. Implicatedness acknowledges both the constructedness of our knowledge about the past as well as an underlying affirmation that inheres in such constructions of the world. We bring our subjectivity and creativity to a model just as much as we cede portions of our subjectivity to it. Unlike the critic in exile, we are implicated in the very structures and networks through which we build our representations of those structures and networks. As I discuss in my conclusion, entanglement is a far more appropriate way of thinking about literary modeling than the disembodied eye in the sky of the lone, distant reader.

Finally, distributed reading suggests a fundamentally relational and reflexive way of thinking about literary meaning, how the sense of a text is always mutually produced through the construction of some context. A context is not introduced solely to explain some text (the means of production, modernity), nor is a single text used to interpret an entire context (*Hamlet* as a window onto the early modern period or the Bonaventure Hotel as a key that unlocks late capitalism).[45] Rather, they bring each other into view as text and context mutually construct one another. In chapter one, I use the notion of the distribution to understand the semantic distinctiveness of highly punctuated poems in the twentieth century, while in chapter four, reading distributions highlights the phenomenological investment of the novel when compared to nonfiction. In chapter five, I explore the distinctive semantic profiles of female protagonists of novels written by nineteenth-century women authors as well as this profile's afterlife in the science fiction novel of the late twentieth century. In each case, the distribution helps us understand the ways in which language signals, how it *distinguishes*, within some larger population of works.

At the same time, distributed reading is also far more spatial in nature. In place of the sentence as the organizing principle of text, one that has been operative from Augustine's bibliographic conversion ("as I came to the end of the sentence . . .") to Gertrude Stein's modernist reformulation ("Sentences not only words but sentences and always sentences have been Gertrude Stein's life long passion"),[46] distributed reading attends to a fundamental dimensionality of texts. As I try to show in chapter two, computational models offer new ways of thinking about the spatializing nature of literary emplotment, while in chapter three I delve into the past to illustrate how topic models reconstruct long-standing theories about the spatial significance of language. Distributional models move us from what we might call a grammatical theory of reading, where the linear accrual of meaning is hierarchically organized, but also fundamentally open ended, to a more diagrammatical theory of reading in Deleuze's sense of the term, where the space of reading is multidirectional, yet formally bounded. It foregrounds the idea of *composition* as a core principle of literary meaning.

I want to conclude with one final image, a photograph by the artist Idris Khan, which is entitled *Thus Spake Zarathustra . . . after Friedrich Nietzsche* (2007). In it we see all of the pages of Nietzsche's *Zarathustra* superimposed onto a single open book, producing an illegible blur across both recto and verso. There is a deep black smudge that runs down the book's middle or spine, familiar to anyone who has ever tried to photocopy a book, reminding us of the book's now lost three-dimensionality. As Garrett Stewart writes about Khan's work, "Content vanishes before us in a single smudged rush—between their own endpapers—in the simultaneous moment of inauguration and closure, all in an impacted apparition of shutter speed."[47] Khan's work stands in many ways as a complement to Morrissey's static, yet fluctuating, page with which I began. Where Morrissey wanted us to see the page's remarkable ability to persist despite all of its changes (via Joyce), here in Khan we see how the single representation of the page—imagining all pages of a book together—results in an impenetrable blur (via Nietzsche). In a book, all of the words cannot be seen at the same time. They take time.

On the one hand, Khan's work draws attention to the often overlooked visual dimensions of books and reading, what the German philosopher Sybille Krämer has called *Schriftbildlichkeit* (the visuality of writing) or Garrett Stewart "the look of reading."[48] Visuality is often what we look through when we read, much like quantity. But Khan's photograph also draws attention to an impasse, to something we cannot see. The "composite" image shows us the limits of using either photography or the book to understand the book (or the photograph) as a totality. It shows us the limits of the technology of the facsimile. Instead, it gestures toward a different kind of visual composition, one that is more abstract, and yet no less referential, which we saw in Deleuze's prioritization of the diagram.[49] Unlike the facsimile's indexical relationship to the text, the diagram has a more aggregate or compositional nature. As a form, it facilitates the act of understanding books taken together.

Khan's abstraction thus marks out an end of a particular tradition, in which the technologies of the book and the photograph have been used as the exclusive tools of understanding those very same media. It reminds us that visuality does not simply sit alongside quantity as two equally forgotten dimensions of reading (iconoclasm as arithmophobia's twin). Instead, it points to the diagram as a necessary vehicle that can be used to *envision* quantity, to grasp, in however mediated a fashion, the quantitative dimensions of texts. Like the *dia-* of "dialogue," which etymologically suggests the connection of words between different speakers, the *diagram* is a drawing that connects two different sign systems. It draws together. In place of the haptic totality of the book—the way its physical availability underwrites its imagined mental graspability—computational reading relies instead on the perspectival totality of the diagram. The diagram facilitates a reading experience of contingent comprehension, a paratextual form that draws together disparate things to understand them in a conditional way. Visuality in this sense is not another latent dimension of the page, like quantity. It is the medium through which we translate between letters and numbers. *Enumerations* is, in this sense, a deeply diagrammatic book, in both its illustrations and its readings. It is one more sign of the spirit of translation that hangs over this work.

ONE

Punctuation (Opposition)

"Like a ,,,,,,,,,,, this look between us."   JORIE GRAHAM

While writing existed long before punctuation was invented, there is no more rudimentary form of inscription than the punctuation mark. The dot, the line, the curve—these are writing's elements. As marks of punctuation—as period, comma, hyphen, parenthesis, or question mark—they both interrupt and conjoin. They divide, but also mark time. Punctuation makes us feel writing. It makes the virtual real.

There is no shortage of scholarship on punctuation. We have numerous accounts of the invention, the fashionability, and the fall of certain types of punctuation marks (whither the semicolon).[1] And since the late eighteenth century, we also have numerous prescriptive works on punctuation's rules.[2] With the spread of literacy and the expansion of print in the nineteenth century, the manual of style would emerge as a quintessentially modern genre, books of syntactical paternalism encircling the unruly hordes of the printed masses. And then there is the seemingly endless parade of interpretive engagements with singular moments of punctuation: the disputed semicolon at the close of *Faust 2*; the famous dash of Kleist's "Marquise von O."; the missing period at the end of Whitman's first version of "Song of Myself"; or the parenthesis in e. e. cummings's "windows go orange in the slowly" that is both its own line and a visual index of the quarter moon about which the poem speaks—a form of onomatographia, when we use punctuation marks to look like objects in the world.

Throughout all of this, however, we have never had a history of what Georges Bataille might have called the general economy of punctuation, a study of the norms and excesses of punctuation in a given period. What is the meaning of punctuation's distributions, its luxuriant overaccumulation, as well as its rhythmic rise and fall, "the delay of language," in poet Amiri Baraka's words? To study the economy of punctuation, and not just a few singular auratic marks, is to study the way spacing and pacing make meaning on the page. It is to understand the way tactics of interruption, delay, rhythm, periodicity, and stoppage are all essential ways of communicating within literature's long history. The economy of punctuation allows us to see the social norms surrounding how we feel about the discontinuities of what we want to say.

Take for example Paul Scheerbart's *Lesabéndio*, a science fiction novel written in the opening decades of the twentieth century. The story concerns the main character's wish to build a tower to transcend his planetary limits. His goal, he says, is to commune with "the Larger [das Größere]." Quantity for Scheerbart is the new Babel. The planet on which Lesabéndio lives, an asteroid off Jupiter named Pallas, is populated by stretchy people with suction feet who have telescopic eyes. They don't have sex and are born from nuts. Their books are microfilms that they wear around their necks. When they die, they are absorbed by another member of the planet, who stretches extremely high and takes in the dying member through his or her pores. It was one of Walter Benjamin's favorite novels.

Whatever else it is, *Lesabéndio* is unique in its predilection for periods. It belongs to a select group of novels in the German canon that use an almost equal ratio of periods to commas (the average since the late eighteenth century is closer to 2.6 commas for every period).[3] Even more telling is when the novel uses periods. There is a moment around the midpoint of the novel when the amount of periods increases significantly (fig. 1.1). Even the lowest occasions of period use after this moment are above the highest in the novel's first half. What has happened?[4]

Over the course of the first half of the story, Lesabéndio has been building support for his tower. In the process, he has overcome one obstacle after another. But a crisis is reached when his colleague Peka, the artist, feels that his role in the project has been undermined. In the segment with more periods than any other in the novel, Peka cries: "You have destroyed me! You have taken everything from me. You have annihilated me. Your cursed tower has made a poor end of my artistic dreams." It is at this point that Peka begins crying, only to realize that his

CHAPTER ONE

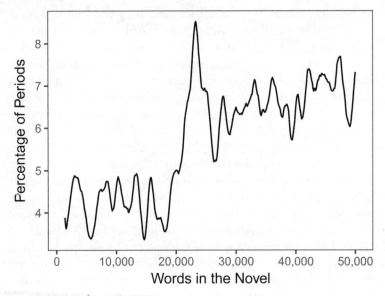

1.1   Rolling mean of percentage of periods across sliding 1,000-word windows of the novel *Lesabéndio*.

tears provide a new kind of glue that is needed to overcome what initially appeared as an insurmountable technical obstacle. Technology triumphs on the fluid surfaces of human sadness.

This moment marks a major turning point in the novel, a kind of conversional axis within the narrative. And it is the period and its accumulation that captures this conversion, the fate of art in the age of industrialization. "This is no longer an artistic story—it is something other—something incomprehensible," says Lesabéndio just after the novel's meridian. The period's rise marks a turning point toward something unknown, something greater than oneself, something potentially inhuman.

The General Economy of Punctuation

Lesabéndio's predilection for periods was not unique. Over the course of the twentieth century, both novels and poetry, at least in English, were increasing the frequencies with which they used periods, at the same time that commas were decreasing (fig. 1.2). Indeed, in the case of English-language poetry, punctuation itself has been decreasing for the past century or so. Poetry appears to be heading toward writing's unpunctuated origins as a form of continuous script.[5]

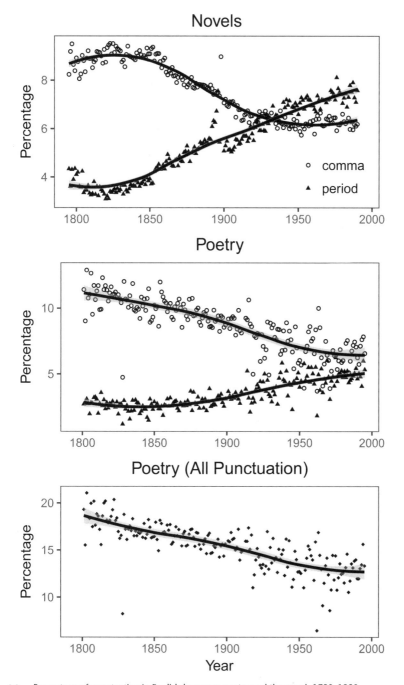

1.2 Percentage of punctuation in English-language poetry and the novel, 1790–1990.

CHAPTER ONE

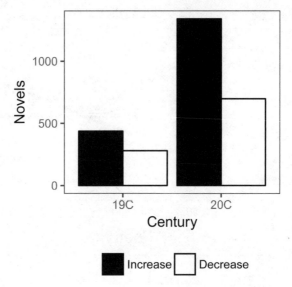

1.3 Novels in English that show an increase or decrease of periods over the course of their narratives in the nineteenth and twentieth centuries.

If novels like *Lesabéndio* were increasingly relying on periods to mark their punctuatedness, this was true within novels as well. Like *Lesabéndio*, more often than not novels in English tend toward using more periods as the plot progresses, especially as we move into the twentieth century (fig. 1.3).⁶

One way we might try to understand this is to see it as a trend toward narrative resolution. The pensiveness and contradictions of commas give way to the clarity and pointedness of periods. Periods mark ends, and thus there are more of them as the narration reaches an end. This is certainly true in a work like *The Sorrows of Young Werther* (1774), whose commas decline precipitously toward the close of the novel as we move out of the young man's sentimental outpourings and into the colder, more clinical narrative of the editor. The utter erasure of punctuation from Molly Bloom's monologue at the close of James Joyce's *Ulysses* tells the same story in reverse. The absence of punctuation reminds us how much this novel resists closure.

But rather than see the period as a mark that exclusively indicates an end, whether narrative or glottal, Scheerbart's novel suggests the power of the increased quantity of periods to signal a sense of an opening. Quantity makes a difference to the dot's meaning. As Lesabéndio ascends his tower ever further into the cosmos, we do not move toward resolution, either in the visual or plotted sense; rather, we move, like Faust, outward and upward, into increasing degrees of abstraction and the "incomprehensible." The punctuatedness of periods, counterintuitively, initiates a profound sense of openness.

This chapter is an exploration of the general economy of literary punctuation. For Bataille, whose work grew out of the soil of French surrealism, the secret of life lay in its superabundance, the fundamental fact of increase that lay behind it.[7] "General economy" did not for Bataille signal a closed system of inputs and outputs, a form of circularity or homeostasis; rather, the economic was far more a model of expenditure and excess. Life was too much. It necessitated that "glorious operation," which Bataille labeled "useless consumption." It turned the rationality of production on its head. The project of art was no longer to make us aware of the productive forces of society, that familiar Marxist credo. Instead, its aim was to draw attention to the necessity of waste and excess as conditions of a full sense of being, of feeling more than ourselves. Art made the fact of excess recognizable; this was the sense of life's "accursed share." For Bataille, this is what allows us to feel our way into something beyond ourselves.

Strange as it may seem, computation may be one of the more effective ways to study this idea of general economy that Bataille had in mind. Computation allows us to see the overall distribution of literary features and identify those spaces of either lack or excess that give shape to a given genre or period of time (or even the notion of the "time period"). Far from a handmaiden of empiricism or the "rational economy" that Bataille wanted us to move away from, computation enables us to inhabit, and more importantly *see*, that accursed share of writing, those spaces of aesthetic expenditure and luxury that offered keys to understanding human beings for Bataille. Reading for quantitative excess follows in the direct path of Bataille and his early twentieth-century surrealist roots.

Such thinking, it should be pointed out, deviates strongly from the norms of statistical reasoning. Outliers are traditionally seen as problems, the exceptions that help prove a rule. Bataille's interest in excess was undoubtedly influenced by the rise of statistical thinking in the nineteenth century, with its strong emphasis on norms and normal distributions. As Wilhelm Lexis, one of the pioneers of statistical thinking in Germany, wrote in 1877, "The state of a human community is on the one hand partially determined by the positive historical forms and norms of both society and state . . . ; but it is also determined by the common and relatively constant actions and afflictions of individuals in diverse settings, which in their discrete units cannot be comprehended but which produce characteristic mass phenomena [*Massenerscheinungen*] accessible to scientific observation."⁸ The Bataillean point of view, on the other hand, suggests a more dialectical relationship between norms and excess, the ways in which the act of exceeding one kind of norm can produce its own form of normative behavior, that accursed share that so fascinated Bataille.

My focus in this chapter will be on the relationship between punctuation's excess and its manifestation in twentieth-century poetry. Few narratives are more strongly ingrained in the field of poetics than the growing antipathy to punctuation in the twentieth century. From modernist sound experiments to the later depunctuated work of poets like William Carlos Williams, punctuation in the twentieth century is most often thought to bar access to the acoustic and rhythmic immediacy of poetry. As Marinetti writes in his "Supplement to the Technical Manifesto," establishing a basic motif around which subsequent work would organize itself: "Words freed of punctuation shine on each other, interweaving their diverse magnetism, following the uninterrupted dynamism of thought."⁹

And yet, at the same time as this allergy to punctuation grows, we can also observe one particular type of punctuation, the period, become increasingly deployed. The period arguably becomes twentieth-century poetry's accursed share. Far from enacting Bataille's dream of dissolution, from clearing a space of freedom and release, the period's excess seems to capture more of a sense of irresolution and antinomy. As I will try to show, the period's abundance brings us into a language space marked not only by a sense of the elementary—more deictic and rudimentary—but also by that of opposition and conjunction, a sense of the irreconcilable. The ending, as Scheerbart's Lesabéndio intuited, is also an opening.

Using a collection of 75,000 poems written in English by 452 poets who were active during the twentieth century (POETRY_20C), this chapter explores poems that deploy periods well in excess of the norms of their age. In doing so, it asks what they might have in common in addition to their punctuatedness, how one kind of excess might establish other kinds of expressive norms. This is what I meant above about the exploratory nature of this chapter—it does not start with a clear, demarcated hypothesis that it sets out to test, but with a more general sense of a scholarly narrative about twentieth-century poetry's antipathy to punctuation that it sets out to complement. But it is also exploratory in the sense that it starts at the beginning of things, using perhaps the simplest computational technique there is, "grep" (which stands for "**g**lobally search a **r**egular **e**xpression and **p**rint"), to extract the frequency of the simplest typographic symbol there is, the period, in order to understand something about its distribution and meaning.

As we will see, the spaces of over-periodization that computation helps bring into view cut more transversally through the traditional ways poets have been divided up in the twentieth century by either period or school. While over-periodization does become distinctly racialized as a poetic practice—a means of traversing a sense of self that can neither cleave itself off nor fully commune with a larger population—such racialization only tells part of the period's abundance. African American writers, for example, are 2.8 times more likely to be represented in the high-period group than they are in the collection at large.[10] Nevertheless, they still comprise only 17% of all poems in the high-period group. There is a stylistic and communicative diversity to these poems, one that revolves around the more general question of too many endings ("over-ending"), of what comes after the end. How do we think and dwell in this excessively punctuated moment?

The Excessive Period

What does the economy of the period look like in twentieth-century Anglophone poetry? Where is the space of its excess and what does it say? The first step in answering these questions is to extract all periods in poems and calculate their ratio relative to the number of words in a given poem.[11] From there, we can then understand the ways in which the period's frequency is distributed across the collection of twentieth-century poems. Figure 1.4 presents a histogram that shows the distribution of periods expressed as a percentage of words per poem across the entire collection. A histogram is a useful tool to visualize how a particular value is distributed within a particular data set. Here the x-axis refers to the percentage of periods in a poem relative to the total number of words in that poem (so in a 100-word poem, 5% means that there are five periods detected in that poem). Each bar in the graph represents a single percent, starting at 0. The y-axis tells us how many poems have that value. The numbers of poems with values beyond 20% are unfortunately so few that they cannot be seen here. The graph has also been artificially cut at 50% to make it more legible though, as we will see, it extends out to 165%, the maximum amount of periods per words in our data set.

What this plot and the data behind it show us is that a majority of poems fall within a very narrow range. Fifty percent of poems have a

CHAPTER ONE

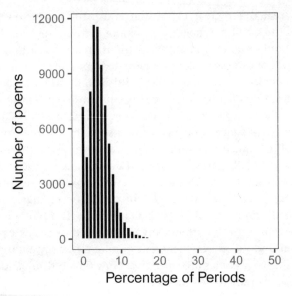

1.4 Percentage of periods per poem in twentieth-century poetry

rate of between 2.35 and 5.88 periods for every hundred words. Ninety percent of the data can be accounted for with a period rate of just under 10% of a poem's words. There is a significant cohort of about 6,700 poems, around 9% of the entire collection, that have no periods at all. Finally, there is a very small cohort of 929 poems, or just over 1.2% of all poems, that use periods to an exceptionally high degree (i.e., more than 3 standard deviations above the mean). This is punctuation's 1%. The range in this group is quite wide, from about 14%, or 1 period for every 7–8 words, to 50%, or 1 period for every 2 words, all the way up to 165%, in which there are more periods than words, as in Amiri Baraka's "American Ecstasy":

AMERICAN ECSTASY (1934)
Loss of Life or Both Hands or Both Eyes. . . . . The Principal Sum
Loss of One Hand and One Foot. . . . . The Principal Sum
Loss of One Hand and One Eye or One Foot and One Eye. . . . . . . The Principal Sum
Loss of One Hand or One Foot. . . . . One half The Principal Sum
Loss of One Eye. . . . . One fourth The Principal Sum

The histogram can help us better understand the distributional nature of punctuation in twentieth-century poetry, showing us the range of its norms and the areas of its excess. What it cannot tell us, of course, is what periods *say*, in either their normal or excessive states. What are the semantic associations that accompany periods in poetry? For this we need other kinds of methods that are able to think about the relationship between the distributions of periods and the distributions of language.

One way to do this is to look at all of the words in poems that are followed by a period and rank them by their increased likelihood of occurring next to a period (fig. 1.5).[12] This gives us a "sense of an ending," in Paul Muldoon's words.[13] What we see are the very pronounced ways that poems tend to end their sentences with an emphasis on death and departure, but also on nature's most expansive features: the *sea, sun, sky, air, world,* and *God*. Periods bifurcate between marking a strong sense of closure, the absolute end of things, and a turn to the most expansive spaces imaginable. They imagine the *again* of the end.

A second approach could look at only those poems from the long tail of the distribution in figure 1.4, at those Batailléan moments of excess. When poems deploy periods in excessive amounts, are there norms or commonalities of expression that surround such excessive punctuatedness? What are the other more semantically inflected features that distinguish these poems from the rest?

Before proceeding, it is worth pausing over two crucial concepts that this last sentence introduces—"feature" and "distinguish." What is a *feature* of a poem? The short answer is anything! Anything that can be drawn from the literary critical tool kit that can be modeled can be a feature. Words, phonemes, rhyme, meter, word count, line length, the presence

1.5   Words that appear significantly more often followed by a period. Size indicates the increased likelihood of ending a sentence relative to the word's overall occurrence.

of rhetorical devices, ambiguity, synonymy, hypernymy, vocabulary richness, syntactic irregularity, difficulty, topoi—any of these categories can be features, and many of them will be used over the course of this book. Indeed, it cannot be emphasized enough just how important this area is for future research. Our ability to create features that accord with the kinds of things we care about as readers is a crucial dimension underlying the successful analysis of literature in a computational setting.

Once we have features, what does it mean for something to be *distinctive?* Here again, there are many ways that we can assess whether some feature is significantly different for one group compared to another. These can include using the notion of the "standard deviation," invented by Karl Pearson at the end of the nineteenth century, which I used above (somewhat problematically, as the data is not normally distributed!). It can take the form of comparing two distributions and making an estimate about whether the differences in those distributions appear to represent two different populations (t-test or rank-sum test). One can look at likelihood ratios, where we compare whether the rate at which something occurs in one group is significantly higher or lower when compared with the rate of occurrence in another group (including a Fisher's exact test or Dunning's log-likelihood test, where the former was already used above on the words distinctive of ending sentences). Or finally, one can test whether a particular value is higher than one would expect by chance by randomly shuffling (permuting) the data and seeing how often that value might occur under different circumstances (randomization test). All of these methods will be used throughout this book at different times, because they each rest on different assumptions and tell us different things about our data.

But what they all have in common is the attempt to estimate whether something we have observed is "significantly" different from some set of reference values, that is, the difference is not just an effect of random fluctuation but *signifies*, at least quantitatively, to readers that something different is happening. Throughout, I will be using the threshold of $p < 0.05$ as my cutoff of statistical significance, though it will not be treated as binary and in many cases will be adjusted lower to accommodate the problem of multiple testing. An open question for the field is the extent to which statistical significance and semiotic significance overlap, one that I will be looking at closely over the course of this book.

To return now to the question of distinctive features, I begin with three of the more straightforward features currently available: words, parts of speech, and hypernyms. In keeping with the building-block nature of the book, I want to start with the simplest proxies I can think of for the lexical, syntactical, and conceptual dimensions of these poems. While the first two might be self-evident to readers, the third requires a bit of explanation. Hypernyms are what linguists refer to as the more generic terms of more particular instances (the hyponyms). *Furniture*, for example, is a hypernym of the word *chair*. The resource WordNet, a lexical database developed by linguists at Princeton University that now consists of numerous mirrors in other languages, has a useful architecture in which every noun is located within a larger tree of hypernyms. "Chair," for example, looks like this:

chair → furniture → furnishing → instrumentation → artifact → whole → object → physical entity → **entity**

Each higher word is intended to be a more general representation of the one(s) below it, culminating in "entity" (where all nouns are entities). I will have more to say in chapter three about the *limits* of hierarchical trees for representing meaning (as does Umberto Eco). For now, I am interested in using these representations as a way of capturing the larger conceptual focus of poems, where "conceptual focus" is being modeled here by the hypernymic categories to which nouns belong. Thus, for every noun in a poem, I transform that noun into the full inherited hypernym tree from WordNet.

Like all models, this has its limitations: not all words will be correctly identified as nouns; not all nouns will be contained in WordNet; concepts are not reducible to nouns; and not all hypernyms will be reasonable approximations of the conceptual focus that I am after. Nevertheless, this approach gives us one way of beginning to generalize from the individual words into larger categories or concepts (we will see a second example in chapter three through the use of topic modeling, and in chapter four through the lexicon-based approach of LIWC).

Table A.1 in the appendix presents a portion of the results of this exercise, showing the features that poems with excessive amounts of periods use in greater as well as lesser than expected amounts.[14] (The full tables upon which these calculations are based are contained in the supplementary material.)[15] These are the features that make these poems *distinct*, through either their presence or their absence.

Reading over this table, we can begin to develop insights about the stylistic orientations of these poems. We see a greater prevalence of personal pronouns at work here, one that is, however, not equally distributed among all pronouns. *I* and *you* are far more common in high-period poems, while *he* and *his* appear less than expected. There is an interesting

CHAPTER ONE

sense of implication to this writing (*I am, you are*, but also *there is*) that establishes a foundational sense of self and other. At the same time, language comes across as more elementary—there are fewer adjectives, adverbs, determiners, prepositions, and conjunctions (though the overall length of words does not shrink).[16] Language feels shorn of its excess, more urgent, but also potentially more general. There is an increase in the present tense, but also a greater use of "a" rather than "the" (as in "a man," "a woman," "a shadow," "a mirror," "a dream"), as well as plural rather than singular nouns. At the same time, looking at the hypernyms, we see how these poems also seem to emphasize communication and connection. Generality here passes through indefinite objects rather than quantities and qualities. There is a hardness to these poems, but one that is also tied, as we will see, to the softness of bodies, to passing boundaries. The emphasis on communication is not tied to specifies (from A to B), but to indefiniteness and plurality, an unspecifiable longing to connect without knowing how or with whom.

Such physical contradiction is complemented by the syntax of negation through the abundance of the lexeme "no" as well as the category of "negation."[17] But it is also marked by transformation (the hypernym "change" stands out as well). These poems tend to negate more, but also call things into question. As we will see, at times this has to do with race, identifying with "black" and what it means ("black" is one of the most distinctive words of these poems, though not shown in the table). At other times it can be even more general than that. "Black" is not always a sign of race.

Ultimately this bird's-eye view can only tell us so much. We have to move into the poems in order to understand the way these words and features are inflected with particular meanings. But these results serve as an important means of establishing an interpretive framework, a medium that allows us to move between the extreme subjectivity of interpretation (*I see this*) and the accountability of our observations to something external to ourselves (this is what is *there*). Reading computationally is the amalgamation of this bifocal process, a way of reading oppositionally, if you will, a mode that is suggestively being operationalized in the high-period poems of the twentieth century.

Consider the opening lines of the Scottish poet Edwin Morgan's "One drinks paraquat" (1990), written at the age of seventy and one of the most densely populated poems with periods in the twentieth century:

One drinks paraquat. One drowns with bricks on.
One skids on black ice. One is strangled with nylons.

One is in hypothermia. One has thrombosis.
A lion crunches one. A lover poisons one.
One is shot at Entebbe. One meets a shark.
One hangs from barbed wire. One is battered with hammers.

Here we see the repeated presentness of verb tense combined with the deictic ambiguity of naming. *One drinks, one drowns, one skids, one is, one has.* The speaker points, and yet we do not know to whom. One, not I. The polarity that the one implies—one is not me—is mirrored in the way every line symmetrically contains two periods. This poem is "periodic" in multiple senses of the word. The period is both a break, a hard stop, and a caesura, a pause within a line that mimetically continues to repeat itself. The period marks a moment of catching one's breath, but also a conceptual division. It links, but it also tells the story of one death after another, much like Ted Hughes's highly perioded macabre poem "Examination at the Womb-Door": "Who owns these scrawny little feet? *Death.* / Who owns this bristly scorched-looking face? *Death.* / Who owns these still-working lungs? *Death.*" The concatenation of death is the period's paradox.

Morgan's "one" and Hughes's "death" are signs of the repetitive ends that surround periods, signs of repetition's endingness. In Angela Jackson's "a beginning for new beginnings," on the other hand, we see the hopefulness that gets attached to the period's breaks:

and some where distantly
there is an answer
as surely as this breath
half hangs befo my face
and some where
there is a move meant.
as certain as the wind
arrives and departs
from me.
and always.

Angela Jackson is one of the most celebrated poets in contemporary American culture and of course a powerful voice for African American experience. Here we see her beginning *in medias res* with the conjunction "and," the period's antithesis, but also concluding with the fragment "and always." The period's overpopulation is also what makes possible the dream of perpetuity, the "and always." Breaks signal struggle, as in the beautiful line "i. and. we. struggle.," but they also signal potentiality,

CHAPTER ONE

as in the spatial break in the stanza above between "move" and "meant." That subtle space allows us to move from the embodied act of "movement" to the mental act of what a "move meant," of making meaning through time. Periods remind us to look closely at these breaks (like the broken "befo" that tells us what or who this poem is really about in its dialect), where the space makes all the difference.

One of the poets whose work appears frequently within the class of poems that use periods in significantly greater numbers is the African American poet Amiri Baraka (1934–2014), one generation ahead of Jackson. In his "An Agony. As Now." (1964), unlike Jackson, who begins in the middle of things, we see how for Baraka the period infects even the title. Pain and time merge, but are also kept apart. In Baraka's hands, the period marks a feeling of time, a feeling that suffers from an absence of continuity. "Now" and "Agony" are the poem's poles. They are linked, but not entirely. The poem opens with an invocation of being contained, one of the more common ways from our hypernym list that highly punctuated poems express themselves.

I am inside someone
who hates me. I look
out from his eyes. Smell
what fouled tunes come in
to his breath. Love his
wretched women.

On the inside, the I is double. "I am inside someone who hates me." But unlike Morgan's parallels of periods—two per line—we can see how Baraka undoes this tight coupling between periods and line breaks. Something hangs on after the period. Most often, it is words of sensing (look, smell, love). Feeling and interiority are at stake here; they are what remain after the period. But they are also things that need to be pried apart.

(Inside his books, his fingers. They
are withered yellow flowers and were never
beautiful.)

The fingers are inside the book rather than on the outside grasping it, much like those words that are inside parentheses. Hand and book are inverted, just as the hands are said to be withered and "never / beautiful." The line break reminds us to pause over the absoluteness of the claim of their non-beauty. Maybe once or someday in the future they will be. Beauty hangs on through time. Like books, periods contain, but never entirely (also like books).

The poem concludes with a definition of what the speaker lives inside:

a bony skeleton
you recognize as words or simple feeling.
But it has no feeling.
[. . .]
It burns the thing
inside it. And that thing
screams.

The speaker wants us to be careful about personifying things, about lending things feeling too easily. You may want to recognize a skeleton as words or feeling, "but it has no feeling." Periods remind us not to engage in these easy attachments, to see the thingness of things. In the end, however, the thing is endowed with feeling, reminding us that it (he) is more than a thing. And yet the scream is alone, cut off from the thing. Nothing is more punctuated than a scream. It cries out and makes itself heard, and in the process cuts everything off from itself.

Periods in poems very often do this job of containing, of burrowing and burying. But as in Baraka, they never do so entirely. This is their strangeness. Commas, on the other hand, seem to calmly partition. They tend to accentuate pensiveness and reflection, not disjunction. "Returning upon my poems, considering, lingering along," writes Whitman in one of the most comma-filled poems in English ("As I ponder'd in silence"). Or, as he writes elsewhere in the final edition of *Leaves of Grass*, "A vast similitude interlocks all, / All spheres, grown, ungrown, small, large, suns, moons, planets, comets, asteroids." The comma is the vehicle of interlocking, of the "all." As Cecil Alexander writes in his highly comma-ed "Looking Up" (1882): "Still upward, upward, everything." Around the comma, totality and the vertical sublime conjoin.

The comma allows us to linger, to imagine parts as wholes. This is what seems to recede in the twentieth-century highly perioded poem. The period is concerned instead with contradiction, with the both/and. In Jackie Kay's "Landslide" (1993), we hear (emphasis on sound intended) a compelling version of such combinatory thought:

His body is buried in the land.
You are buried here: this place
where your voice is

CHAPTER ONE

a disused mine. You mime
the same sentence. The sound gets stuck.
Life. Life.
You wait for the time that never comes.

His body is buried in the land, but yours is here. Here where? In the poem or in the world? To be buried in a poem leaves open the possibility of finding one's way out from underneath the words. After all, the poem is described not as a tomb, but as a "disused mine." Challenging, but ascendable. The specificity of "this place," which does not tell us where we are, however, suggests otherwise.

Perhaps even more interesting is what happens in the stanza that concludes with the periodic staccato "Life. Life." We open with the image of the "disused mine," a metaphor for the buried voice from the previous stanza (not to mention the master metaphor of data analysis). But this "mine" then turns to "mime" ("You mime the same sentence"). Instead of *mining* the sentence for its meaning, the speaker can only *mime* it—one's voice, what is *mine* in another sense, is lost. As the poem says, "The sound gets stuck."

And yet the sound has told us everything we need to know here: the same sound means two different things (mine/mine), just as the consonant shift (mine/mime) indicates how even the smallest acoustic difference changes everything in a poem, especially by removing us from the world of sound and into that of the mime (the ultimate acoustic shift, that is, into non-sound). It is a point nicely made in Anthony Barnett's highly periodied "Memory" (one of the most highly represented authors in this group), where he writes, "You do not want. / Unite / And Untie." Even the slightest shifts make a difference, as that "i" migrates around the "t," especially when it comes to remembering.

But in Kay the poem insists that the acoustic evacuation embodied in the mime can still be heard: the n/m shift between *mine* and *mime* will be replayed in reverse as a form of chiasmus between *same* and *sentence*. The repetition implied by the word *same*, which replays the *mime*'s double *m*, will be transformed into a sequential thing called a *sentence* (opening forward rather than downward as with the mine). But sentences too can be constraining, not just through their syntax and their dependencies, but in what they can say. As with the death sentence, sentences are forms of judgment. They often close things off.

This playful movement between down and out, between repetition and resonance, will find an echo in the poem's subsequent concern with the return that does not happen. "You wait for the time that never comes,"

Kay continues. Or later: "You buried something and forgot / where you put it. Years ago." Did she bury it years ago or forget years ago? Not just the thing, but even the memory of the thing is lost. Kay's punctuated "Years ago" is like a rejoinder to Baraka's "As Now." There is no immediacy here. "Remembering is dying slowly," she says a second time. Is the act of remembering dying, or is the act of remembering *like* dying? And what does it mean to say it *twice?* The refrain, that which keeps coming back, tells us about what is lost without letting us know exactly what it is.

These are just some of the ways that when we read these poems more closely we can begin to "flesh out" what those lists of significant features are telling us—to see how the period, in its most excessive moments, can be understood to tell a particular story, a story about the desire for doubleness, not wholeness. There are of course numerous other ways the period could mean or not mean. It would be pushing it to say that there is some kind of fundamental message that these poems communicate. I want to be open to the multiple ways these poems signify even as they share a particular, and particularly excessive, feature. But I also want to see the ways in which they cohere, the commonalities that run through, if not all of them, a significant number to make them look and feel different as a group. Our readings, framed by the quantitative, can lead us to more quantitative ways of testing and inquiring that can help us generalize about the group as a whole, which can in turn frame our rereadings of the poems. Number is a lens.

By way of example, I want to conclude with one final model in order to test whether this sense of semantic doubleness holds more generally across these poems. I want to put into practice that circular model of reading that I drew on in my introduction (fig. 0.1), where the close readings that are used to validate distant readings also generate new kinds of models for further distant readings. This is another way of being implicated in our models.

In reading these poems, I felt that the period marked a kind of hinge, one that conjoins rather than divides, that allows us to experience a sense of antinomy, to hold two opposed things or ideas or even sounds in our heads at the same time. One way we might model this sense of hingedness is to use the vector space model that I discussed in the introduction to measure the semantic relatedness of the words on either side of a period. Does the excess of periods correlate with a greater sense of semantic disjunction across the boundaries of a sentence?

How would this work in practice? Using a word-embedding model like word2vec, we can model the semantic similarity of every word to every other word in our collection of poetry.[18] For example, according to such a model, the words most similar to "stone" in twentieth-century poetry are *rock, slab, broken, hewn, wall, granite, crumbling,* and *marble*. Using an NLP tool for detecting sentences and subsetting by sentences that only end in periods, we can combine the vectors for each word in a sentence together to generate a single semantic representation of the sentence as a whole. We can then traverse a poem by moving from one sentence to another and compare the similarity of each sentence to the next one. Because sentences can in some cases be very long, I focus only on the final clause prior to a period and the first clause after a period, where there are clauses.[19] The greater the dissimilarity between sentences or clauses on either side of a period, the more semantic disjunction we can argue there is surrounding the use of a period (fig. 1.6).

Using this model, I found that high-period poems, while they do not exhibit significantly greater amounts of semantic diversity overall, do exhibit a significantly higher standard deviation within an individual poem, though the effect size of this difference is not large.[20] What this means is that when there are multiple sentences in a poem, the range of semantic difference between the sequential pairs is somewhat greater than in a random selection of poems taken from the twentieth century at large. While the poems in aggregate are not enacting more semantic difference, they are enacting slightly more difference *within* the poem itself. It suggests that these poems oscillate slightly more strongly between sentences that are more similar and less similar, a tactic that might help foreground the sense of contrast that I felt when I read them. One way of explaining the feeling of semantic difference that one experiences in these poems is that it occurs because of the greater *contrast* of similarity and difference across different sentences in the same poem.

Taking all of these observations together—the distinctive features, the readings, and the vector space model—it seems fair to say that the pe-

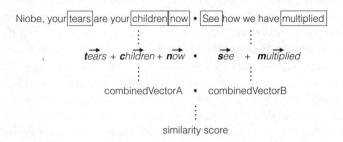

1.6 Schematic representation of a model to compare the semantic similarity of sentences across periods.

riod, in its excessive poetic state, does appear to have a particular kind of meaning associated with it. It asks us to look both ways, to be in the moment but also think more generally, to move from the very similar to the very dissimilar, to hold a paradox in our heads without clear ends. This seems to be the point of Kenneth McClane's "42 Ways of Looking at the Harlem Hudson Line" (1981), another high-period poem that uses the period to articulate a space of ambiguous social connection within the train car, a sense of movement that also isn't:

The man next to you speaks to you.
The man next to him speaks to him.
Everyone is speaking.
You feel like speaking.
You speak.
Nothing has changed.

The period is active, present, and communicative. Everyone is speaking. It feels, it conjoins, and it divides all at the same time, and yet seemingly without consequence. Nothing has changed, period. And yet we go on adding more. More is the knowledge of its own failure. This is what too many periods also tells us.

TWO

Plot (Lack)

"Meanwhile the whole history of probabilities is coming to life, starting in the upper left-hand corner, like a sail." JOHN ASHBERY

When Irish monks began separating words in manuscripts by spaces in the seventh century, little could they know that they were performing a central task of computational text analysis. Knowing where words begin and end remains one of the core tasks of any computational approach to reading, and, one could add, to just about all modern approaches to analytical reading.[1] While letters, morphemes, and phonemes all underlie words, it is still the words or lexemes to which we most often return to ground a text's meaning. After punctuation, they are the next most elementary particle of writing. Their accumulation, and their evacuation, provide the scaffolding upon which we intuit a sense of a text's meaning, what Walt Whitman once called the drift of language.

This chapter is not about words in the individual sense—keywords or even particular types of words (pronouns, prepositions, sentimental words). Rather, it is about words in the distributional sense, as a larger set of patterns or behaviors, as a form of *usage*. A great deal of recent scholarship has focused on the history of single words or groups of words (we might call this first-generation computer-assisted reading). Aided by searchable databases, scholars have been mining individual words now for quite some time.[2] One of the elementary insights of distributional semantics, however, is that the frequencies of *sets* of words convey a more inclusive and nuanced sense of meaning to readers or listeners. Rather than home in on a single luminous word, distributional semantics encourages us to think about the relationships that exist between words and how meaning is shaped by such probabilistic distributions. As words recur, we begin to develop mental models of what to expect as we read.[3] If only 5% of Jane Austen's *Pride and Prejudice* consists of unique words, then a vast majority of the novel's significance is contained in the repetitions that run through the fabric of the text. Understanding those repetitions as probability distributions, as a set of likelihoods, allows us to understand what a text is conveying at a more formal or semantic level.

The argument that I want to make in this chapter is that understanding texts as word distributions gives us an important new way of thinking about plot, the way actions and beliefs are encoded in narrative form. Comparing word distributions allows us to think about the shift or drift of language in a text and the way such transformations signal to readers a change in the text's concerns. In this, I am drawing on recent work by Matthew Jockers and others that has refocused our attention on the foundational narratological concerns with the distinction between narrative discourse (*sujet*) and the narrative's content or story (*fabula*).[4] As Gérard Genette suggested in his foundational work on this subject close to half a century ago, "It is surprising that until now the theory of narrative has been so little concerned with the problems of narrative enunciating, concentrating almost all its attention on the statement and its contents."[5] Much early computational work in this area has followed suit, focusing more on topics, motifs, or content than the question of narrative discourse. New work in the field is interested in asking instead how larger formal structures—the organization, not the specific content, of language—convey meaning beyond any specific motif or topic (the quest narrative, marriage plot, man bites dog, etc.).

Where Genette's thinking about narrative discourse was largely inspired by the adjacent field of cinema studies (in his notions of diegesis and frame), much of the current work today on emplotment takes its cues from drama. The "arc" or trajectory is currently the most common model through which plot is being understood, as a theatrical theory of denouement is mapped onto prose narrative.[6] The most popular proxy for such plot arcs has to date been the use of sentiment analysis to model emotional intensity as a sign of narrative turning points. The arc serves as a marker of narrative tension and resolution, translating Aristotelian catharsis into novelistic terms.

My work in this chapter will depart from these existing models by focusing on the more general question of linguistic drift, the ways in which

CHAPTER TWO

narration marks out moments of linguistic change beyond a sense of emotional tension. Rather than think in terms of things rising and falling (and back again), I am interested in understanding a sense of semantic contraction (or expansion), a sense of expressive heterogeneity that is either lost or found. How do narratives open up to more linguistic diversity as they progress or, conversely, close down into a narrower space of expression? And what might these changes have to tell us about a text's meaning, about the narration of change itself?

For narrative theorists, I hope my questions will resonate with existing models more closely than has perhaps been the case up until now. Transformation, contraction, shift—these are all elementary models of narrative change that, as we will see, are grounded in the history of narrative theory. The computational models that I am working out in this chapter are drawn in various ways from the long history of thinking about narrative form. At the same time, the aim of this chapter is to draw attention to new kinds of narrative forms, to highlight a set of common practices across a subset of novels that have so far escaped the attention of traditional critical models. In focusing on the shifting distributions of words, computation brings into view important new literary communities.

To think more closely about the question of plot, I will be focusing on a particular group of novels that are distinctive in their investment in a sense of lexical contraction. By this, I mean that the distributions of language that characterize these novels become markedly more similar to each other over the course of the narrative's progression. The works I will be discussing are drawn from a group of 450 largely canonical novels in three languages (German, French, and English) that have been curated by my lab (NOVEL450). What makes these novels unique is how the linguistic "space" within them contracts in ways that are distinctive when compared to the vast majority of other novels. And yet, as we will also see, understanding that contraction is by no means a straightforward process. This chapter is about slowing us down from too quickly labeling things and proving things and spending more time understanding what these patterns might mean. It is about moving from plot as a noun to plot as a verb, from thinking about "a plot" in novels to thinking about what it means "to plot" as a form of transformation. In many ways, it is a broader reflection on the meaning of emplotment as a core way of thinking about literary significance.

Novels have, for much of their history, been imagined as an abundant form. They are often exceedingly long (think of Cervantes's *Don Quixote*, Grimmelshausen's *Simplicissimus*, Scudery's *Artamène*—for me, at least, they never seem to end). They also multiply dramatically over time. Already by the end of the eighteenth century, readers felt there were far too many. Today, there definitely are. Classical theories of the novel have in turn mirrored this typographical excess. "The composition of the novel is a paradoxical fusion of heterogeneous and discrete elements into a perpetually renounced organicism," writes Lukács in his *Theory of the Novel*.[7] Bakhtin's "heteroglossia" could be seen in this sense as nothing more than the theoretical outcome of the history of the novel's imagined excess—novels always say more than other kinds of texts. More words, whether as types or tokens, mean more meaning. Novels, so we have come to assume, have never been lacking.

The novels on display here are interesting precisely because of the way they push against this history of the novel's imagined excess. For these novels, such as Goethe's *Wilhelm Meisters Wanderjahre* or Joachim Campe's *Robinson der Jüngere*, Virginia Woolf's *Mrs. Dalloway* or Jane Austen's *Sense and Sensibility*, Jules Verne's *De la terre à la lune* or Lamartine's *Le Tailleur de pierres*, the narration of some kind of lack, the *loss* of polyphony, serves as one of the narrative's driving concerns. These are novels that are interested in exploring a form of social constraint as it is experienced through language. If the poetry of denotative surplus that we saw in the previous chapter comes together to think about questions of resolution—of how more could ever unify to become true—the art of lack, I would argue, offers us insights into what it means to contract inward, and in so doing potentially say more. The art of lack is the dream of insight where there is increasingly less and less to say.

Lacking Novels (Model 1)

To begin, I want to return to Augustine's *Confessions*, which first emerged in my introduction. Augustine's work provides a useful point of reference for computational reading because of the way it explicitly establishes the practice of repetition as a cornerstone of personal transformation. For Augustine, reading plus time equals change. Doing something over again allows us a new way of accessing a different aspect of ourselves.

But the *Confessions* is important in another way as well, one that has more narratological implications. The *Confessions* was historically unique because of the way it inscribed the notion of personal transformation as a core purpose of narration. The idea that conversion should form the foundation of the individual life, that life should be defined by a singular turn that is at once a turning away (from some former self) and a turning back (to one's true self), marked a major historical departure from more classical conceptions of the self and its possible narration.[8] Such personal rupture was seen not only as a vehicle of return, the con- of *conversio*, but also one of belief and commitment, of a turning *toward*. The strong binary form of the text, the sense of a before and after, was framed by Augustine as a way of generating a devotional stance in both the protagonist and reader. It provided a model for the production of readerly attachment.

For later theorists such as Genette and Todorov, the purpose of narration was largely related to the logic of causality. Narration for these thinkers allows for the causal concatenation of events (first A happens, then B happens as a consequence of A, etc.). Unlike this more elementary serial structure of narration, Augustinian conversion introduced a different kind of template in which narration was marked by a strong sense of before and after, by a sense of difference.[9] This is another way that reading plus time produces change. Now, however, it is the time of reading itself that matters, rather than the time *between* reading.

We can actually observe this effect at work in Augustine if we take a distributional approach to modeling the *Confessions*, once again using a vector space model to spatially represent this sense of a narrative turning point. Treating each chapter (or book, as it is called in Augustine) as its own vocabulary distribution, we can compare these distributions to each other to get a sense of the work's overall narrative structure. To do so, we divide each book of the *Confessions* into a separate document and transform it into a vector of word frequencies. After removing stopwords and words that appear in fewer than half of all the books, the beginnings of such a table would look like this:

	abs	absit	adhuc	adversus	aetate	aetatis	...
Book 1	6	0	5	2	1	1	
Book 2	5	0	1	0	0	3	
...							

We can then calculate the similarity of each book's vector to every other book's by using a vector space model that I discussed in the introduction, where the frequencies are treated as spatial dimensions.[10] Here I use Euclidean distance, which plots each book in as many dimensions as there are words in the model (in this case, 479 dimensions). The dis-

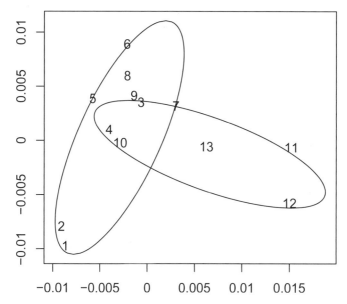

2.1 Semantic relationships between the thirteen books of Augustine's *Confessions*. Ellipses have been drawn around the pre- and post-conversional books.

tance between any two books becomes a proxy for their semantic similarity. The further two books are from each other, the less their vocabulary distributions resemble each other, and the less they are thought to be "about" the same thing.

Figure 2.1 plots these semantic relationships between books using the process of multidimensional scaling (MDS), which reduces the multidimensional relationships between books down into two dimensions so they can be visually represented on the page. I have also artificially added ellipses to encompass the pre- and post-conversional books so we can observe their spatial clustering. While some information is inevitably lost as we move from 479 to two dimensions, the graph offers an approximation of the semantic similarity of each book to one another.

What we see happening in this graph is a twofold process in which the later books of the *Confessions* become increasingly dissimilar to the earlier ones and also increasingly dissimilar to each other as they begin to pull away along a different axis, especially books 11–13.[11] Augustine not only speaks in very different terms before and after his conversion, but he also speaks *increasingly* differently. According to this graph, conversion is an entry into not only discursive change, but also discursive expansion. As Augustine scholar James O'Donnell writes about the

CHAPTER TWO

post-conversional books, "What Augustine learned to do at Ostia he now *does*, in writing this text. This is no longer an account of something that happened somewhere else some time ago; the text itself becomes the ascent. The text no longer narrates mystical experience, it becomes itself a mystical experience."[12]

Augustine's work thus gives us a useful model for understanding the narrative process of conversion, one that we can implement using the tools of vector space models and distributional semantics. While I have explored the question of narrative conversion at greater length elsewhere—of when a text opens itself up to novel ways of speaking[13]— what I want to focus on here is the inverse question of when a narrative *closes down*. What happens when a novel inverts this Augustinian paradigm, when it enters into a space of discursive contraction rather than expansion?

Table A.2 in the appendix presents a list of the most contractive novels by language.[14] Among these we find, in English, Ann Radcliffe's *Mysteries of Udolpho*, James Joyce's *Ulysses*, and George Eliot's *The Mill on the Floss*, Jules Verne's *De la terre à la lune*, Jules Renard's *Poil de Carotte*, and Louis Duranty's *Le malheur d'Henriette Gerard* in French; and Theodor Fontane's *Irrungen, Wirrungen*, K. P. Moritz's *Anton Reiser*, and Paul Scheerbart's *Lesabéndio* in German. These are novels that strongly contract their range of expression when compared to a larger population of novels. Theodor Fontane's *Irrungen, Wirrungen* (1887) provides a useful example of what this process of contraction looks like for the most strongly contractive novels (fig. ? ?).

How might we better understand what this sense of contraction means? One way to do so is to inspect the vocabulary that is shifting between the first and second halves.[15] If there is a reduction, what words are appearing more in the second half when compared with the first? What kind of vocabulary accompanies this semantics of contraction?

When we do so, we see a rather interesting, and potentially problematic, insight. In almost all cases of highly contractive novels, there is a significant shift in the use of proper names. This can either take the form of a change in the name of the main character—for example, in *Lesabéndio* the narrator increasingly refers to the main character as "Lesa" (which sounds like "Leser" or "reader"). Or it can take the shape of different characters playing more of a role in the latter half of a novel. In Joyce's *Ulysses*, for example, it is "Stephen" and "Bloom," while in Fontane it is "Käthe" and "Franke," the substitute spouses for the novel's initial lovers, "Lene" and "Botho."

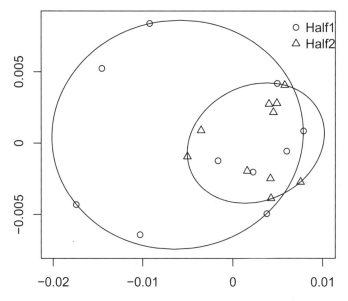

2.2 Semantic relationships between the first and second halves of Fontane's *Irrungen, Wirrungen* (1887).

As we will see in chapter five, proper names comprise a huge amount of words in a given novel (an estimated 4% of all tokens, or ca. 4,000 occurrences in an average-length novel). A shift in the use of a few names can significantly alter the overall spatial relationships between chapters. And, as we will see in the next trial of this model, removing proper names significantly changes the outcomes of the test. In one sense, this poses a problem because we are not seeing vocabulary shift as much as we are seeing some kind of shift in the *social network* of the narrative. And yet this too is decidedly interesting. When a character is no longer referred to as "Anton" but instead as "Reiser" (as in Karl Philip Moritz's biographical novel *Anton Reiser*), we know something crucial about him has changed. Those shifts are often accompanied by other significant changes in vocabulary as well. In the case of *Anton Reiser*, it is an indication that he has matured into an adult, leaving behind a world of *books, learning, God*, and his *parents* for one shaped by abstractions such as *life, being, man, fate* and, crucially, *loneliness*. The proper name captures a larger shift in personhood toward a life full of doubt and metaphysical uncertainty.

In the case of Theodor Fontane's *Irrungen, Wirrungen*, the social network itself conveys a great deal of information about the nature of the novel (fig. 2.3).[16] Fontane depicts an affair between the aristocratic Baron Botho and the working-class Lene, who will split roughly halfway through the novel and marry more appropriately to their respective stations, preserving the class hierarchies of Wilhelmine Berlin. Despite the successful resolution of the social square, the novel was received by early readers as a scandalous work of fiction (referred to by one reviewer as Fontane's "whore's tale").[17] The social network is generated by observing every unique time two characters (or locations) appear on the same page together and drawing an edge between them when they do so. Figure 2.3 presents the social network for each quarter of the novel, where the thickness of the edges indicates an increased number of co-occurrences between entities.

As we can see as we move through the novel in quartiles, there is an initial cluster in the network where Botho and Lene are enmeshed via the two older women, Frau Dörr and Frau Nimptsch, comprising the domestic space of the ironically named "Schloß" (Q1). By the second quarter of the novel, we see Botho's social world expand (all those names on the right) as Lene gradually moves away from the protected space of the castle (Q2). In the third section, we see Botho caught between the two lovers, Lene and Käthe, as well as the two competing spaces of the castle and the balcony (Q3). Finally, in the last quarter, we see how the realignment has taken place, with Botho/Käthe paired off and aligned with the space of the balcony, while the working-class couple, Lene and Franke, move together (Q4). Both the garden and the castle—the novel's two idylls—retreat from narrative significance by the novel's end.

Fontane's work is almost always prized for the dialogical richness of his writing, a linguistic diversity that is achieved through his novels' attention to dialect and class.[18] According to critics, *Irrungen, Wirrungen* is one of the ultimate nineteenth-century exemplars of Bakhtinian heteroglossia. And yet that changing social network that leads to a sense of the novel's lexical contraction tells a different story, one in which such polyphony is not universal within the novel, but has a distinct trajectory. Heteroglossia is not the point of Fontane's social realism. It is the experience that we leave behind as we enter into social institutions like marriage.

More importantly, I would argue, is the way this lexical contraction is counteracted by a different way of thinking about language in the novel, where a sense of the diversity of language and character in the novel's first half is replaced by its semantic equivalent, as polyphony gives way to polysemy. The doubleness of space that marks the novel's geographic structure (the two poles or "worlds" of Botho's and Lene's domestic settings between the garden castle and the urban balcony) is translated into a linguistic model of language's own doubleness.

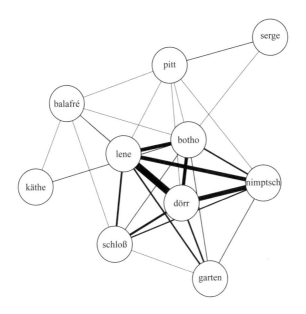

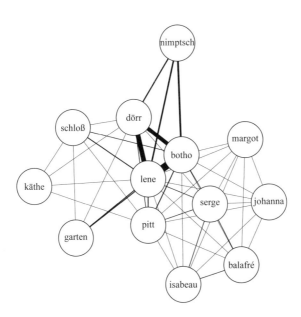

2.3 Social network of Fontane's *Irrungen, Wirrungen* divided into quarters.

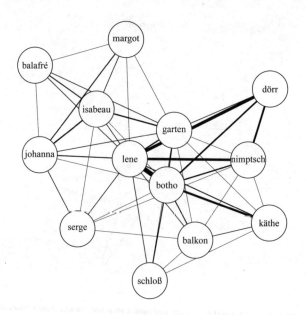

Q3

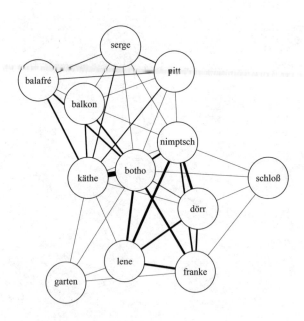

Q4

2.3 (*continued*)

This new linguistic world in the novel is perhaps best represented in the figure of Botho's new wife, Käthe, whose role expands considerably in the second half of the novel. Unlike Lene, who stands for the crossing of class and dialect, Käthe captures what happens to language when it is put into a social box. While there are numerous examples of this turn, one of the most telling is Käthe's repeated use of the word *komisch*, one of the words unique to the second half of the novel and used far more often here than in German novels more generally. "It is just too funny [*Es ist doch zu komisch*]," "Ah, that is too funny [*Ach, das ist zu komisch*]," "Can you think of something funnier than that? [*Kannst du dir was Komischeres denken*]," or "Love-letters, too funny [*Liebesbriefe, zu komisch*]."

There is of course nothing funny about "love letters," let alone this particular novel. *Komisch* can also mean "strange" in German, and it is this strangeness of the comical that I think Fontane is driving toward. There is a nonliteral aspect to communication that emerges in the latter half of the novel that is associated with the process of socialization, an enfolding of meaning inward as characters move outward. The turn toward social constraint, the socialization of marriage, that marks the novel as a whole is compensated for by a concurrent sense of semiotic expansion. The novel's denotative excesses captured in Fontane's use of dialect and class-marked speech in the novel's opening sections are replaced by the connotative excess of irony by the novel's close.

Such semantic doubling will find its scenic correlate in the space of the balcony that we saw become more central to the novel's spatial structure. This is the place where much of the concluding dialogue between Botho and Käthe transpires. It is also the location that Jonathan Crary has identified as a quintessential liminal space of modernity, enshrined above all in the work of Manet.[19] At the close of the nineteenth century, balconies serve as architectural in-between spaces between public and private, between the domestic interior and the urban exterior. Much like the word *komisch*, urban balconies in the late nineteenth century conjoin two different spaces of meaning. The balcony becomes a sign of the sign's imagined excess.

Fontane's investment in marriage as an experience of both lexical constraint and semiotic expansion gives us one way of thinking about the formal conventions that surround the novel of marriage. We can find a poignant echo of this in one of the classics of modernist writing, James Joyce's *Ulysses*, also one of the more contractive novels in our collection. It is fair to say that few novels have a more rambunctious relationship to lexical richness than *Ulysses* (except of course *Finnegan's Wake*). In order to measure this, we can use a classic calculation such as

CHAPTER TWO

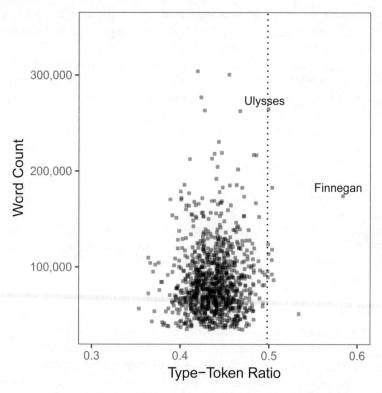

2.4 Average type-token ratio of 1,000 novels published between 1880 and 1930 using 1,000-word samples.

type-token ratio (TTR), which compares the number of individual word types to the overall number of words (or tokens).[20] The values range between 0 and 1, where higher scores equal more "richness," that is, fewer words repeat themselves. A score of 1 would mean that no word ever repeated itself in a passage. Using TTR, we see how Joyce's *Ulysses* does indeed exhibit more vocabulary richness than 99% of novels written within a half century of his (fig. 2.4).[21] Nevertheless, it is also interesting to note that *some* novels do appear to have a slightly higher average vocabulary richness than *Ulysses* (but not *Wake*). These include Louis Tracy's *The Great Mogul* (1900), Louis Stellman's *Port O' Gold* (1922), and Ernest Bramah's unsung *Kai Lung Unrolls His Mat* (1928).

The interesting thing about type-token ratio in Joyce, however, is not that it is so high (which would come as little surprise to readers), but rather that it *declines*. If we plot a rolling average of TTR across individual pages of the novel, we see a precipitous fall toward the novel's close (fig. 2.5). The shift, perhaps needless to point out for Joyceans, happens exactly at the start of the eighteenth chapter of the Paris edition (around page 487 in the graph), when Molly begins her long, unpunctuated monologue. It is at this point that we see a significant decrease in the lexical diversity of the book, dropping 9.7% in a single page and 11.2% across the entire chapter relative to the overall average within the novel.[22] Perhaps most interestingly, as we observe this decline in TTR, we also see a precipitous increase in the synonymy scores for these pages, meaning there are an increased number of words that are synonyms of one another (i.e., share similar semantic contexts).[23] At the very moment when Joyce's vocabulary contracts, the rate of synonymy spikes, increasing 112% in a single page and 33% above Joyce's overall average.

To put this in more familiar terms, with a raw decline of about 6.5% of word types for the final chapter, this means that for every hundred words of text there are on average about 6–7 fewer word types, so that a reader encounters about 30–35 fewer unique words per page. This is no small change, especially when repeated over an entire chapter. At the same time, because there is also about a 5.7% rise in the number of words that are synonymous with one another, a reader similarly encounters about 25–30 more words per page with potentially overlapping semantic contexts.

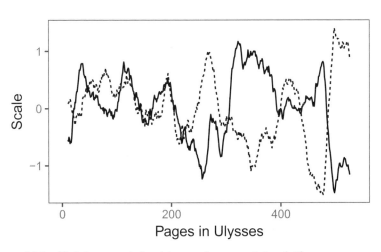

2.5   Relationship between vocabulary richness and synonymy in Joyce's *Ulysses*.

CHAPTER TWO

What this is telling us is that there is a relatively strong correlation between the decline of word types and an increase in synonymy in Joyce ($r = -0.347$, $p < 2.2\text{e-}16$). While the correlation is significant, it is not exceedingly strong, able to account for only about 35% of the variance in the data. It should be pointed out that this effect is not unique to Joyce, but common across writing more generally. If we examine the relationship between 10,000 pages taken at random from the modernist collection, we see a very similar level of correlation ($r = -0.362$) between the two measures.[24] This means that when Joyce's vocabulary becomes simpler in terms of types, it also becomes simpler in terms of the types themselves—the use of fewer rare words and more elementary words increases the overall semantic relatedness on the page. There is an implicit argument being made here about the relationship between marriage, lexical constraint, and semantic surplus. As we move into the interior space of marriage, the linguistic world collapses even as the semantic resonance of that world increases.

This is precisely the argument, we remember, made by Fontane's *Irrungen, Wirrungen*, where the double confusion of the novel's title (it can be roughly translated as "Errors, Confusions") suggests that the relationship between lexical and semantic diversity is not simply one of correlation, but one that has a particular directionality. In the novel of lack, we move *from* lexical diversity *to* the semantic kind, from the polyphony of the novel's first half ("errors") to the polysemy of the novel's second ("confusions"). The phonetic repetitions buried in Fontane's punning title highlight the semantic excesses to come. One gets lost in the excess of denotation (*Irrungen*), while one gets cognitively disoriented by the connotative kind (*Wirrungen*).

The physical waywardness of these novels' protagonists, Baron Botho or Leopold Bloom ("errant" in both senses of the word, not to mention in their B-ness), turns by the end of each of these novels to the semantic disorientations of their wives, Käthe and Molly. The position of being a wife becomes a microcosm of the larger social world, where the lexical reduction that marriage implies—becoming lexically redundant by taking the husband's last name—is followed by a semantic opening up. These novels invest in a sense of social and linguistic constraint to locate a new sense of literary potential.

These are just two examples of the ways that the discursive constraints detected by the model translate into forms of emplotment that revolve around not just a profound sense of change, but one that is accompanied by a combined sense of lexical restriction and connotative surplus. The closing down of a certain dimension of language (lexical richness) corresponds with an opening up of another kind (irony and synonymy). The point is not to argue that these represent universal types, but that novels that correspond to this type of narrative model exhibit other distinctive spatial qualities (in the sense of their emplotment), qualities that tell a particular story about the novel's meaning that is not exclusively tied to its themes or content or a sense of lexical abundance.

If we look at another subgroup of these novels, the works of science fiction that appear in table A.2, we see how these novels too are interestingly marked by a basic narrative trajectory that heads toward a state of incommunicability. Verne's *De la terre à la lune*, for example, is not only essentially a novel about ballistics—about arcs, anticipating the later Pynchonian reverie of *Gravity's Rainbow*—it is also about the impossibility of the return trip. The human inhabitants of the rocket never actually land on the moon. Confirmation of their arrival, of the entire success of the novel, is never achieved. The circle of conversion remains essentially unclosed. Similarly, the Time Traveler's narrative in Wells's *Time Machine* is labeled as "fiction" because there is nothing that can be brought back from the future. His listeners remain incredulous in a moving reversal of the novel's origins as a form of credible narrative (from the found manuscript in *Don Quixote* to the Defoeian claims of truth in fiction). Much like the lost protagonists of Verne (or Pynchon), *The Time Machine* will end with the narrator's disappearance, in this case a disappearance into Time itself.

We can see something similar at work in Paul Scheerbart's *Lesabéndio*, where once again the protagonist disappears, this time into outer space. As Lesabéndio ascends into the farther reaches of outer space, an ascent that feels unmistakably Faustian, he will go blind and encounter a series of allegorized voices. He begins in a state of laughter but ends in one of excruciating pain, an experience that is framed as a radical reorientation of the senses. Transcendence of the planetary is rendered as a profound physiological rupture. But it is also finally depicted as silence. "But Lesa said *nothing*," Scheerbart writes (emphasis in original). The "new life" that is invoked in the novel's final sentence—conversion's ultimate trope—will be rendered by Scheerbart in the subjunctive: "And the green sun shone so brightly—as though a new life awoke on it as well" [*Und die grüne Sonne strahlte so hell auf—als wäre auch auf ihr ein neues Leben erwacht*]. At the end of the science fiction novel of lack, we are left with nothing but the conditional, with the uncertainty of transformation. This is the form of semiotic excess that cannot be communicated with the planetary remainder. The transformations of proper names and social networks that create a more general sense of lack in these novels are accompanied by an explicit investment in an aesthetics of incommunicability.

Lacking Novels (Model 2)

These then are some of the ways that the lexical contractions led by a largely nominal association with proper names lead to a sense of expressive restraint. Changes in proper names bring with them strong semantic shifts in the novel that correspond with social factors like marriage or technology within our collection of largely canonical novels. It raises a potentially interesting hypothesis to test on bigger samples of novels. Can we identify this as a signature of these genres more generally? But what if we also wanted to control for proper names and see if the distributional diversity of a novel's language contracts regardless of shifts in character?

Successfully removing all proper names from a novel is a challenging task, and one that I undertake in two successive steps. The first is to use a list of proper names from each language, which inevitably misses very unique literary names (such as Botho or Bloom) as well as very generic names (such as miss, the count, doctor, etc.). To solve this problem, I generate a list of words that are common to a majority of novels in each language and then keep only those words in my analysis. Rare names will not appear in over half of all novels. The downside of this approach is that one loses rare words that may be significant for a given novel. Fontane is a case in point. The Berlin dialect that is so important to the earlier part of his narrative and unique to him as a writer falls away in this more general novelistic language. This approach also requires a good deal of trial and error. Several trials were required before no traces of proper names remained in the high-scoring novels.[25]

When we successfully implement these two steps, however, we find a strongly altered list of novels (table A.3). Some writers are still there, but most have changed. New books include Virginia Woolf's *Mrs. Dalloway*, Jane Austen's *Sense and Sensibility*, Alphonse de Lamartine's *Le Tailleur de pierres*, Joachim Campe's *Robinson der Jüngere*, Goethe's *Wilhelm Meisters Wanderjahre*, and Rilke's *Malte Laurids Brigge*. The concerns raised by novels as they contract also differ considerably from novel to novel. It is not easy to identify overarching concerns that accompany such lexical contraction. But in many ways that is the point. Lexical contraction is independent of particular content—it *is* the content of these novels.

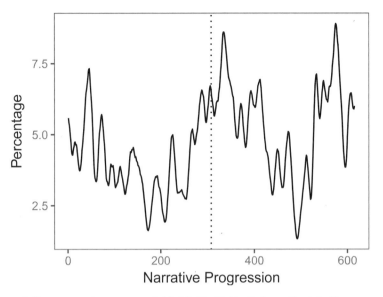

2.6   Rolling average of proper names in Virginia Woolf's *Mrs. Dalloway*. The dotted line represents the novel's midpoint.

Once again we can examine distinctive words for each half of the high-scoring novels to try to understand the particular nature of the semantics of contraction. On the whole, we see very different concerns for each of the novels. For example, in Verne's *De la terre à la lune*, one of the few novels on both lists, we move from a world of quantity and measurement (*cent, mille, quatre, fait*) to one of travel, the heroic individual, and the public crowd (*héros, homme, voyageurs, travers, foule*).[26] The novel shifts dramatically at the very moment that we move from the scientific discussion of the conditions of escape velocity to the question of "manning" the technology (when the brave volunteer, from France of course, emerges from the crowd and offers to captain the missile to the moon). The hero replaces the "club" as the novel's dominant concern. As the construction of the missile turns to the spectacle of its launch, as man is encased in the technological and not the other way around, language shrinks inward.

In Woolf's *Mrs. Dalloway*, we move from the interior world of momentary thought and marriage (*thought, moment, morning, marry*) to the more collective space of the party—*people, ladies, family and friends*. Things get shallower, but also more populated. If we return the proper names to the novel and observe their prevalence as the novel progresses, we see

CHAPTER TWO

a distinctive shift in terms of their increased frequency (though also an increased variance as we see more highs and lows) (fig. 2.6). The second half of the novel has considerably more proper names per 1,000-word windows, culminating in the legendary cocktail party.[27] As people occupy more space, the lexical diversity of the novel contracts. There is less room for the kind of cogitative expansiveness that marks Woolf's fiction more generally.

Readers of *Mrs. Dalloway* will know that as the novel progresses, Clarissa moves from the subject of interior monologue to its object ("For there she was," concludes the novel). Like the unknown car with which the novel opens, Clarissa becomes that symbolic enigma that holds society, or a certain portion of it, together. Like Joyce or Fontane, the position of the wife that is signaled in the novel's title once again foregrounds the lexical contraction of marriage. Alongside the psychological opacity of the "Mrs.," we also encounter the opacity of death. The suicide of Septimus Warren Smith will occur in the second half and hang over the gaiety of the cocktail party like a shadow. "Oh! thought Clarissa, in the middle of my party, here's death, she thought." Death, Woolf later writes, is an attempt to communicate. It is a negation that says something. This is one way to understand the point of Woolf's contractive emplotment.

Woolf will interestingly touch explicitly on the issue of conversion in her novel. As Peter Walsh ruminates on the value of "proportion," he thinks, "But Proportion has a sister . . . Conversion is her name." According to Walsh, conversion is the antipode to proportion, that which undoes equilibrium. It is the force that Mrs. Dalloway and her party both aim to counteract, but in doing so also enact. The very social form that is meant to keep conversion or death at bay brings about the experience of readerly disequilibrium.

Mary Ward's *Robert Elsmere*, one of the more strongly contractive novels in our set, will also use the vocabulary of conversion, less subtly and far more overtly than Virginia Woolf. From the early domestic scenes of young girls outside and in (*girls, young, valley, garden, school, house*), we end with a strong prevalence of religiously inflected language (*love, God, faith, human, heart*). Unlike for Augustine, the devotional turn in Ward means a turning toward a more uniform vocabulary distribution. Life gets more constant and less variable in this conversional space. In German, we see how the Robinson Crusoe rewrite by Joachim Campe moves from a world of assent, parents, nature, and the divine (*ja, Gott, Eltern, Bäume, Sonne, Welt*) to one of savages, objects, weapons, and appearances (*Wilden, Sachen, Waffen, schien*). As the adventure quotient increases, so too does the predictability of the language. Finally, in Goethe's *Wilhelm Meister* sequel, we can see a strong shift from a world of entertainment, family, and religion (*Tante, Alte, Religion, Unterhaltung*) to one dominated by art, craftsmanship, but also business (*Künstler, Hand, Geschäft*). As a novel overwhelmingly concerned with the process of socialization, the "mastery" and integration that is supposed to follow the journeyman years in the guild model of labor, we see the late Goethe arguing against his earlier *Geniepoetik*. Instead, he moves the reader toward an imagined reconciliation between art, handicraft, and commerce. The compromise between art and the market admits its own expressive defeat.

On Emplotment

This chapter has tried to offer an exercise in thinking about plot via vocabulary distributions and, in the process, to show us a group of novels that share an affinity for producing a sense of expressive lack. Whether this takes the form of a change in the social network or an overall reduction in lexical dissimilarity, these novels offer us an experience of what it feels like to have our linguistic universe become more homogeneous as we read. In some cases, as in Fontane or Joyce, this can be associated with an increased sense of irony or polysemy. As the world narrows in some lexically distinct way, a sense of the potentiality inherent in language increases. In others, as in Goethe, the process of socialization is accompanied by a solidifying of categories and associations, as the diversity of practices and subject positions in the novel's first half gives way to the consensus surrounding the triad of art, craft, and commerce in the second, one that we are arguably still arguing over today. In *Mrs. Dalloway*, the move from deep interiors in the novel's first half to the shallow surfaces of the cocktail party in the novel's second becomes a sign of a looming sense of semic evacuation, a death both symbolic and real that resides at the core of modern life for Woolf.

More generally, this chapter has tried to argue for the importance of thinking schematically about narration and the ways in which computational models can be a useful tool for thinking spatially about narrative form—for seeing the form of narrative. While I have offered one new kind of model here, I want to conclude by providing an overview of the larger landscape of existing models for the study of plot as a spur to further research. Each model brings with it assumptions about the schematic nature of narrative discourse and different quantitative implementations of those assumptions. No one model can capture all dimensions of narrative

CHAPTER TWO

Table 2.1 Plot models

Narrative model	Technique	Measure
Expressive Constraint	Vector Space Model	Global Semantic Contraction
Expressive Contrast	Vector Space Model	Global Semantic Difference
Turning Point	Foote Novelty	Local Semantic Difference
Social Interactions	Social Network Analysis	Character Co-Occurrences
Emotional Trajectory	Sentiment Analysis	Rise and Fall of Emotions
Content Arc	Principal Components	Thematic Change
Pacing and Plotlines	Frame Detection	Events Per Time

plot. There is a great deal of future work to be done to think about the possible ways we can capture the spatiality of narrative discourse.

Table 2.1 shows seven different models for measuring narrative change. The first refers to the model I have been working with here that observes the strength of lexical contraction in a novel. But we have also seen how social networks can capture a different understanding of plot, one that is more related to the structural relationships that define a narrative's arrangement. We still do not have a full-length study of how social networks work in novels, a problem that is in part owing to how difficult they are to extract.[28] But it is certain there is a great deal of information encoded in the ways in which character ties are established, undone, and redone in a novel. Alex Woloch's question of the relationship of the one to the many, that is, the protagonist's relationship to the secondary set of characters in a novel, offers one productive way of thinking about social relationships.[29] But one could also focus on the top end of the distribution of characters, to see how these core relationships evolve over narrative time. Do we see strong binary relationships at work, in a theory of the narrative companion (à la *Don Quixote*), or more triangular and transitive structures, as in Jane Austen, or the narrative quadrant, as in Goethe's *Elective Affinities* or Fontane's *Irrungen, Wirrungen*? By introducing the notion of time into the study of literary social networks, these models can give us a sense of narrative structure according to social arrangement rather than semantic distribution.

Two more approaches that utilize vector space models are also available. The first approach looks similar to the one used in this chapter, but rather than identify contraction, we could identify those novels that have the strongest difference between their first and second halves, that are in a sense well sorted in a binary way. These are novels like William Thackeray's *Vanity Fair* or Henry James's *Portrait of a Lady*, where we move from the independence of Isabel's single life, a world of "lovely" things and "hope," to the social strictures and emotional pain of her

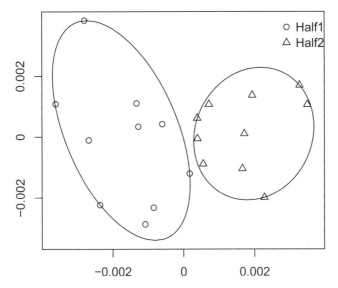

2.7   Semantic relationships between the halves of Henry James's *Portrait of a Lady*.

married life, "loved" in the past tense (fig. 2.7). The novel's semantic binariness conveys, once again, a driving argument about gender and married life.

A second and related approach would look for singular moments of transition, of when a narrative's progression is marked by a profound sense of before and after. Here the concern is less with an overall distinction, an establishment of two opposing worldviews as in James, and more with a sense of breakage, of a dividing *point* (fig. 2.8). For this, we can adapt an approach from the realm of music theory called Foote novelty to think about emplotment, one that tries to identify points of maximum difference within a linear work of art.[30] This method will reappear in more detail in the final chapter when looking at a poet's career. Suffice it to say for now that it allows us to see if there is a significant moment of change that comes to define a narrative (or career) in a unique way. Any value above the dotted line represents levels of semantic change that are seen in less than 5% of all random permutations of the data for a given novel. When run on all of our novels, it turns out that these turning points are overwhelmingly situated in the later parts of a novel.

Two models that have been implemented by Ben Schmidt and Matthew Jockers look for different kinds of semantic trajectories. Where Schmidt focuses on using principal components as a way of identifying the shifting thematic concerns across the narratives of television shows,

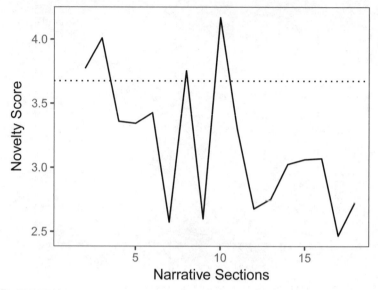

2.8 Inter-chapter novelty in Kafka's *The Castle*. The dotted line represents a level of semantic change that happens in less than 5% of all random permutations.

Matthew Jockers looks at changes in sentiment vocabulary to map a sense of narrative fortune over time.[31] The rise and fall of a vocabulary associated with positive and negative experience is designed to capture the rising and falling fortunes of the narrative's plot.

Finally, I would like to add one more approach to this list that tries to think about narrative change in terms of the idea of the "frame." Drawing on Genette's initial interest in cinematic structure, the frame can be thought of as similar to the "scene" or "shot" in a movie. The goal here is to identify the boundaries of narrative units and then understand the pacing of their segmentation.[32] What constitutes a scene change in a prose narrative? Do these scenes coalesce into distinct plot*lines* within a narrative? And how often are we switching between these local units (frames) and more global units (plotlines) within a novel? While very much a work in progress, this approach allows us to begin to understand the linear assembly of narratives in a more fine-grained way and how those units are assembled into distinct narrative trajectories. The ultimate goal is to begin to model the temporal alignment of these segments to be able to understand Genette's emphasis on narrative anachrony (the way many narratives do not work in a sequential way or, put more technically, the disconnect between a narrative's *fabula* and its *sujet*).

What all of these approaches have in common is a fundamental rethinking about plot away from high-level motifs or content and toward a distributional understanding of language. The relationship of recurrence with change becomes the foundation for thinking about narrative emplotment. Its value is the way it allows for a variety of different implementations, each of which captures a different theoretical understanding of narrative form. There are many plots to the act of emplotment.

THREE

Topoi (Dispersion)

"And the words wander away . . ."
GASTON BACHELARD, *THE POETICS OF REVERIE*

In 1938, on the eve of the Second World War, the German Romance philologist Ernst Robert Curtius initiated a new field of study, which he called "historische Topik." Like the words *Germanistik* or *Romanistik*, *Topik* implied a form of scientific study, in this case the study of the topics of the past. For Curtius, a *topos* was "something anonymous. It flows into the author's quill as a literary reminiscence. It has a temporal and spatial omnipresence, like a visual motif." A topos was a cultural constant, omnipresent, in Curtius's words, one that crucially contained an impersonal element. "It queries literature not according to the drift of literary forms, but the immutability of spiritual content."[1] Topics were greater than the individual author, portraits of an epoch resistant to both time and historical catastrophe. The echoes of the war reverberate like a clear topic within the emerging discipline of *Topik*.

The study of topics has become newly interesting today, resurrected by the increasingly ubiquitous application of a procedure known as "probabilistic topic modeling."[2] Topic modeling is a practice that identifies the likelihood of the co-occurrence of words within particular documents, assigning words to different groups (the topics) and then those topics to different documents. It is based on the assumption of distributional semantics discussed in the introduction. The expectation of a topic model is that a word's meaning is best represented by identifying those words that most often appear near it. A topic is defined as a generalized associational pattern.

The results of topic modeling are importantly predictive rather than descriptive. The presence of a word within a topic or a topic within a document is based on a probability—it is likely that two words from a given topic will appear together, but this is not necessarily the case. This is what accounts for the model's generalizability, its ability to identify potential semantic configurations that are, according to how Curtius thought of topics, beyond the reaches of time. Instead of representing the actual distributions of documents, topic modeling attempts to identify a more limited number of semantic fields that are likely to appear within them.

Topic modeling has proven to be a very successful tool for identifying coherent linguistic categories within large bodies of texts. It has experienced a wide array of applications in studying large-scale cultural phenomena across a number of different disciplines. Within the field of computer science, a great many researchers are experimenting with ways of improving topic modeling depending on the nature of one's questions—looking at topics over time, or across different types of writing, or between languages, to cite just a few examples.[3] Within the field of literary and cultural studies, researchers have begun applying topic models to the study of the history of scholarship, the media, and even the novel.[4]

And yet, despite the growing body of work on topic models, no one has stopped to ask the question "What is a topic?," either in the classical rhetorical sense or in the computational one. If we have this new way of deriving semantic significance from texts at a large scale, how does it fit within the longer philosophical and philological traditions of understanding "topics"? What, in other words, do these lists of words *mean?*

This chapter is about how words coalesce into something more general, more "omnipresent," in Curtius's words, to form those hard-to-define entities that we have traditionally called *topics*. It continues the process of scaling up my own book's concerns by asking how the entities of the previous chapter (words) cohere to produce a more general category like topics. Unlike the notion of emplotment, where words assume meaningful configurations over time, topics are semantic units that are more temporally invariant. (This is certainly how Curtius wanted us to see them.) While topics themselves change with time, as do the ways we talk about topics, the notion of the topic itself is less dependent on time than the spatial configuration of plot. As the name suggests, topics pull us out of time and into the realm of space, shape, and form.

This chapter is in many ways about understanding a more elementary relationship between language, space, and meaning. The word for

CHAPTER THREE

"topics" is derived from the Greek *tópos koinós*, or common place, later translated into Latin as *locus communis*, which would go on to become a core aspect of knowledge organization for centuries.[5] These terms capture a long-standing belief in the connection between location and ideas, the way *configuration* lies at the core of language's meaning.[6] Where the previous chapter aimed at understanding the relationship between distribution and meaning, this chapter is interested in studying "the figure of speech" in a more literal sense, as the spatial relationship between words. Instead of using topic modeling to study something else, I am interested in studying the problem of modeling topics itself.

To do so, I will begin with an overview of the history of thinking about topics, from Aristotle to Renaissance commonplace books to nineteenth-century encyclopedism. Understanding how topic modeling fits within this longer tradition of deriving coherent categories of thought from a surplus of information—where there has always been a surplus from a single human perspective—will help us see how computation has a distinct pre-computational past. In this, it should help topic modeling lose some of its revolutionary mystique. On the surface, topic modeling is nothing more than one more approach to indexing human knowledge.

At the same time, there are also significant differences with how topic modeling positions us in relation to the past, differences that this chapter tries to bring to the foreground. In the second section, I move on to the study of a single topic, hereafter known as Topic 150, derived from a topic model of German novels written over the long nineteenth century. As we will see, Topic 150 concerns the notion of life and death, and turns out to be one of the most trans-historically vital topics produced by the model. As Michel Foucault argued, "life" does indeed seem to emerge as a new "quasi-transcendental" of modern life.[7]

My interest here, however, is not with "life" in all of its nineteenth-century discursive forms. Rather, I am interested in the more limited question of how this topic, if we can call it that, organizes language in long narrative forms that are themselves very often about individual lives. What conceptual or metaphorical concerns does the topos of life and death carry with it through time? How does it establish new kinds of relationships between documents that we have not seen before? To begin, I do nothing more than read the passages that are associated with this topic to better understand the nature and coherence of computational topicality. As I have argued throughout this book, reading is a crucial aspect of "validation," where we begin to understand the logic of computational modeling and the way it orients us as readers. It moves us past a framework of reproducibility to one of understanding.

In doing so, we see how the genre of the epistolary novel emerges at the heart of this topic, one where questions of life and death merge with those concerning art, aesthetic experience, and a sense of immersiveness. The topos of life and death brings with it (translates, we might say, in the literal sense of carrying something over) a concern with the liquid and immersive nature of aesthetic experience. Buried within this more literal constellation of words associated with life and death is a more latent metaphorical worldview about the intersection of the letter and art.

These are the insights that will emerge from my reading of the texts as they are brought together by a particular model. However "true" they may feel to me—in the sense of faithful to what these texts seem to be saying—they are also entirely conditioned by the particularities of the model as well as the data included. Other models and other data sets would invariably produce other kinds of insights (and once we have enough model interpretations, we can then start to study the relationships between kinds of models and the kinds of insights they promote *ad infinitum*). There is a co-constructedness to reading that is every bit as true for computation as it was for the diverse material formats of books. Topic models are far less indexical tools that point to a stable set of transcendental constants and far more interpretive ones that reinsert us into, or *position* us within, a plenitude of expression.

In the chapter's final section, I move to a more quantitatively driven approach that observes the topic in varying states of coherence. Typically we look at topics at their most likely state, observing a topic's presence within documents in which it registers its highest probability. But topics can also be weakly present as well as co-present with other topics. Documents are not exclusively defined by a single topic, as topics can and do reside across all documents in different states of likelihood. Using the approach of multidimensional scaling from the previous chapter—where word distributions are transformed into spatial representations—we can observe a topic's semantic field at different states of likelihood, just as we can observe the relationship between topics within these different states, how they overlap and intersect. Spatial relationships give us insights into the interconnectedness of topics, but also the way they coalesce and disperse. Space provides access to semantic meaning.

Reading topics in this way, or reading topologically, as I would like to call it, is valuable because it allows us to see how ideas cohere, but also fade away. The history of literary criticism is replete with ways of thinking about the relationship between space, shape, and meaning (from the classical attention to figures of speech to the Renaissance focus on poetics and genre to the more recent "spatial turn").[8] Reading topologically provides a new way of attending to the form of language, this time through an attention to the latent quantities of words. It allows us to envision how figure and concentration serve as an essential foundation of human thought, and that their opposites, dispersion and formlessness, are equally essential for the process of intellectual change. Rather than get caught up in debates about a topic's coherence or validity—the tautological question of when a topic is a topic—modeling topics in this way allows us to see the uncertain boundaries and interspersions that define topical space. Topological reading makes visible the way topics are neither firmly bounded objects stable through time, the transcendentals of philosophical thought, nor clearly evolving genealogical units, the elements of *Begriffsgeschichte* that move coherently from one form to another across linear time. Instead, the ideas of topical space appear more as stages or screens upon which we see the dispersion and the fluidity of linguistic relations accumulate and stretch out according to quantitative performances. They offer a new way of thinking about the "figure of speech" as a core dimension of literary meaning.

Studying topics in this way allows us to see how topics ultimately contain a sense of their own otherness, that, like the computational topics used to model them, each topic contains within itself the potentiality of all other topics in the topical space. Instead of a clear taxonomy or genealogy—the horizontal table of relations on the one hand, the explicit lines of descent on the other—topic models teach us to recognize the differences that reside within sameness, the alterities of thought. As we will see, computational topics are far more heterotopical by nature. Their value lies not in their indexical nature—their ability to comfortably index a stable set of ideas—but far more in the way they overturn the notion of the topos as a "commonplace," the way they help us instead discover *uncommon*places for understanding.

A History of Topical Thought

The theory of topics begins, as with so much else, with Aristotle.[9] For Aristotle, a "topic" was not a coherent or fixed entity, like an idea or a concept. Rather, it was more akin to a formal process, a set of rules through which to derive novel arguments. Its association with a sense of place or topos served as both a mnemotechnical device and also a tool of derivation, to assist in what Aristotle called the process of enumeration: "For just as in the art of remembering, the mere mention of the places instantly makes us recall the things, so these will make us more apt at deductions through looking to these defined premises in order of enumeration."[10] The topic's spatiality is what made it, perhaps somewhat paradoxically, a tool for deductive reasoning, for the linear process of what he calls "enumeration."

As Ann Moss has recounted in her work on the history of commonplace books, the *locus communis* of antiquity would reemerge by the turn of the sixteenth century as a ubiquitous form of knowledge organization.[11] For northern European scholars like Albrecht von Eyb, Rudolph Agricola, and Erasmus of Rotterdam, *loci* or topics provided an essential way of ordering the world. Far from serving a purely decorative purpose, they supplied the building blocks of knowledge. As Agricola would write in *De Inventione*,

> Hence, when we turn to consider any matter in our minds, by following these [loci] we may survey the entire nature, parts, compatibilities and incompatibilities of a thing, and may draw thence an argument suitable to the matters proposed. These common notions, therefore, because they contain within themselves both all that can be said about each thing and every possible argument, men have called places, because in them as in a receptacle or treasury of sorts repose all the instruments for producing belief. A place therefore is nothing else but a certain shared characteristic of a thing, by whose instruction can be discovered what is convincing in each thing. Let the place therefore be so defined by us.[12]

Or, as Edmundus Richerius wrote over a century later in *Obstetrix animorum* (1600), "From this definition it is clear that the term 'commonplace' is to be extended to all branches of knowledge and even to the mechanical arts, and not, as in rhetorical theory, kept within the narrow confines of demonstrative, deliberative and forensic oratory."[13]

And yet, however much topics were aligned within an epistemological framework, they were also never fully divorced from expression, from the production of new forms of speech. In the humanist context, they were designed above all else to be used within social settings. "This [system] has the double advantage," writes Erasmus in *De copia*, "of fixing what you have read more firmly in your mind and getting you into the habit of using the riches supplied by your reading."[14] The emphasis on usefulness and relevance would gradually underwrite the distinction between what Ann Moss has called "proof by argument" and "proof by example." The stringent systems of dialectical reasoning inherited from the medieval period would be replaced by an emphasis on the particular as a means of accessing more universal truths. Over time, topic labels would move away from the more abstract hierarchical systems of Mirabellius's

Polyanthea (1503) or Agricola's *De Inventione*, where the latter begins with the primary duality of *alios internos/externos* from which follow *insubstantia, circumstantia, cognata, applicita, accidentia,* and *repugnanita*. By the time of Erasmus, we see a more horizontally structured system of what Erasmus calls "the things of most prominence in human affairs," a system premised above all else on likeness and antinomy ("they should be arranged according to similars and opposites").[15]

As most historians agree, this system would gradually fall apart over the course of the next two centuries. Individuals did not suddenly discontinue the art of commonplacing after 1800, recording quotations in their notebooks or thinking about the categories of knowledge. But the practice no longer retained its pedagogical and epistemological validity in quite the same way.[16] The critical systems given birth to by Kant, the emphasis on observational truth emphasized by Locke, the premium on originality within a new commercial writing system, and finally the value of synthetic forms of knowledge over the exemplary all gradually replaced the topic or *loci* as the center of modern thought. In Goethe's *The Man of Fifty* (1816), for example, we see how the commonplace system becomes the object of a highly entertaining parody as the protagonist repeatedly disturbs the free flow of social conversation with ill-timed citations of classics. The bedrock of early modern knowledge had, by the early nineteenth century, become a habitual joke.[17]

Such quotation-based models of the commonplace would be replaced in the eighteenth century by new systems of indexing knowledge at the document level, just as the open-ended lists that accompanied topics would be replaced by new forms of compressed, keyword-driven forms such as the modern encyclopedia, which would emerge in the nineteenth century as a staple of the publishing industry. "Amitié" could still be a topical keyword, but in the *Encyclopédie* of Diderot and D'Alembert it was no longer composed of a list of quotations, but a condensed definition of the thing itself: "the pure exchange of the spirit is simply called *acquaintance*; the exchange where the heart is concerned is called *friendship*" [le pur commerce de l'esprit s'appelle simplement *connoissance*; le commerce où le coeur s'intéresse par l'agrément qu'il en tire, est *amitié*].[18] As Chad Wellmon has shown, the growth of the eighteenth-century subject index performed the opposite task, not of condensing knowledge within a digestible format for the common reader, but of pointing the reader outward to an ever-growing archive of documents within an increasingly complex system of "cross-reference."[19] As Ann Moss has argued, "retrievability" became the primary focus of the spatial arrangement of ideas.[20]

In this sense, topic modeling can be seen as a natural extension of the document-level system of indexing that became increasingly popular during and after the eighteenth century (and that had its early modern and medieval precursors). The method was initially designed as a form of "information retrieval," aligning computational topics with the idea of knowledge orientation. Topics help us know where we are in a larger sea or field of information (pick your metaphor). They are useful for the act of partitioning or subsetting groups of documents or even passages into different categories.

In another sense, however, computational topics appear to exhibit a great deal of novelty with respect to the history of topics. Like the subject index, the classical topic is defined by a sense of semantic uniqueness, *distinctio*, a core early concept in understanding the coherence of the commonplace as a distinct place. As Erasmus advised, topics were best thought of as antonyms—Faith and Faithlessness, Reverence and Irreverence, and so on. Difference was one of the core features of their mnemonic usefulness.

The computational topic, by contrast, allows for the identification of a different kind of distinction, what we might call the difference of the same. There are multiple ways, for example, that the topos of the "heart" assumes meaning in the poetry and prose of the nineteenth century, differences that computational topic models aim to separate out from one another. If we run a topic model on a collection of poetry and novels from the period, we can find several topics in which the word *heart* is operative: *tear,* **heart***, sorrow, grief* (Topic 5), *love,* **heart***, kiss, tender, passion* (Topic 46), *dread, fear,* **heart***, suffer* (Topic 48), and *happy, life, hope,* **heart***, joy* (Topic 96).[21] Here the keyword, or what early modern writers would have called the headword or *rerum capita*, is capable of expressing several different, and often conflicting, human emotions, from sorrow to fear to love to happiness. Instead of emphasizing the differences between topics (between Reverence and Irreverence), computational topics are more attuned to identifying the distinctions between different facets of a single topic or topoi.

Early modern systems of commonplacing were not immune to this kind of differentiation either. But they did so in the name of some greater unity. The heterogeneity of statements that fell within a topic was subsumed by the unity of the category to which they belonged. It was precisely this diversity that allowed topics to be so agile in their application, to be useful in different settings across both time and space, whether it was the individual lifetime or the longer time span of historical epochs. This metonymical distance—the gap between the whole for which

CHAPTER THREE

a statement stood and the particular nature of any given statement—served as the basis of the topic's performative openness.

At the same time, the disambiguation between topics in topic models is complemented by a greater degree of ambiguity *within* topics. Classical topics were defined as a one-to-many problem. A single word or concept stood for a potentially limitless archive of citations. Classical topics were open objects, in so far as one could never rule out the possibility of finding one more statement relevant to a given topic. Each addition would in turn alter the understanding of the whole. Indeed, that was part of their pleasure, as well as their intellectual value. The discovery of new sentences or phrases or even words relevant to a topic changed one's overall understanding of that topic. The classical topic was not a static container, but resembled more a dynamic feedback loop. It evolved over time as it grew.

The computational topic, by contrast, incorporates that openness within itself. Rather than group statements under a single keyword or phrase, it organizes a heterogeneity of statements under a complex semantic *field*. It operates according to the principle of many to many. Unlike the indexical certainty of the classical topic or modern subject heading—this keyword points to that phrase, or this heading points to that document—the computational topic is doubly open ended. Something points somewhere, but those two entities of the what and the where are definitionally open and contingent. While we can say which words of a given topic are represented in a document or passage, each document will activate different combinations of words with differing intensities. The "topic" is not a fixed entity traveling through documents, but differently embodied in each of its instances.

This semiotic openness of the computational topic—what a given subset of words points to—also has a relational aspect, one more dimension of the topic's openness. The classical topic was formally open, but categorically closed. One could always add one more statement, but only to reaffirm the system of differences that underpinned topics (like Reverence and Irreverence). One more statement was not going to make one topic begin to look like another. Computational topics, by contrast, operate in reverse: they are formally closed and categorically open. All documents and all words are contained within every topic, just as all topics are contained in every single document. There is a monism to topic models that is a crucial aspect of their identity. Somewhere in Topic 150 lies Topic 73. And within every document, however improbable, exists the possibility of every single topic. It is the ultimate epigenetic system. This is what I meant earlier when I said that computational topics are fundamentally heterotopical. Each topic contains within it the seeds of every other topic.

For scholars of commonplaces like Moss, the decline of commonplacing was to be understood in a tragic vein. The rise of indexing and retrievability, along with the priority of personal possession (this is my language), engendered both a distance to the past as well as a disassociation of writing from collective modes of thought. The commonplace book was nothing more than a personal archive, a "treasure trove" or, in nineteenth-century terms, a collection of "beauties." Citation became a means of promoting a more distinct sense of authorial charisma.

Topic modeling poses a distinct, and to my way of thinking, promising challenge to this largely indexical system that we have inherited. If we move away from the notion of retrievability and toward one of discovery, we can begin to engage with the dual sense of both estrangement and immersion that topological reading makes possible. Rather than continue to point to authoritative sources in a citational system designed to preserve a sense of individual distinction, topic models are far closer to Agricola's notion of a combinatory totality from which new knowledge can be generated. Computational topoi open us up to and immerse us within the heterogeneities of thought, the differences that reside within zones of similarity (and the similarities that reside within those differences). Far from separating and authorizing, computational topics de-distinguish. They allow us to sense what Gilles Deleuze called in his book on Foucault "the relation of the non-relation,"[22] to my mind still one of the most effective ways of capturing the distinct kind of positivism that is topological reading.

Reading Topologically

Topic 150

Leben, ewig, Tod, Welt, Mensch, sterben, mehr, Seele, Erde, Geist, Gott, Ruhe, tief, Nacht, Himmel, Zeit, nie, Gedanke, Stille, sehen

life, eternal, death, world, man, die, more, soul, earth, spirit,
God, rest, deep, night, heaven, time, never, thought, stillness, see

In what follows, I explore a single topic that was generated on a collection of 14,888 passages drawn from a collection of 150 German novels published over a century and a half (1770–1930).[23] My aim here is not model optimization,[24] nor the summarization of an entire epoch, but to

CHAPTER THREE

read a single model to gain a better understanding of the intertextual relationships that topic models can engender. The insights gleaned are thus entirely contingent upon the particularities of this one model. But they are also a first step in beginning to understand the ways in which different model parameters generate different kinds of relationships between texts. Before we start using topics, we need to read a lot of them first.

So why Topic 150? First, because I care about it. We are always in some sense operating within the framework of matters of concern, in Bruno Latour's sense.[25] Topic 150 is interesting to me because it captures an elementary sense of what it means to be human. Between life and death there lies the hope of the eternal. At its base, it provides a refrain of antinomies: life/death, world/heaven, time/eternal, earth/spirit, thought/stillness. Opposition serves as the driver of the topic's cultural duration, antithesis one mode through which the novel as a genre, and the epistolary novel in particular, appears to assume coherence over time.

But I can also motivate this choice through a variety of quantitative features that help me better contextualize the topic within the model. These measures can illustrate what makes this topic stand out as a topic beyond just its distinctive content. Before we zoom in on that alluringly novel arrangement of words, we need to understand its relationship to all of the other arrangements of words.

Table 3.1 provides a list of initial features that we can use to assess a particular topic (one could of course devise many more).[26] Each score is ranked according to which percentile it belongs to for all topics, so that the higher the number, the stronger that score is for the topic. Going down the list in order, the first measure (# tokens) tells us how many overall words are accounted for by the top twenty word types in the topic. This gives us a sense of how strong the topic is within the collection. Is it a dominant, middling, or weak topic in the sense of its lexical presence? Before we make claims about how important something is, we need to understand how frequent it is. The second row (# passages) tells us how many passages have this topic present above some artificial probability threshold, that is, the number of passages where this topic is strongly present. Remember that technically speaking, all topics are contained within all documents, just with vanishingly small probabilities. In this case I will be using the threshold of 20% throughout this chapter, which other researchers have identified as a reasonable approximation of a topic's presence when reading similar-sized passages.[27] Below this amount, it begins to feel questionable whether a topic is "explicitly" present in a document. Seen in this way, such a measure gives us further perspective on the question of the topic's "presence." Because numerous passages come from the same document (each novel is broken down into 1,000-word passages), it is important to understand how broadly represented this topic is across the larger documents rather than just at the passage level (# novels). When fewer novels contain a topic, it tells us the extent to which it can be considered to be specific to a particular work or author.

The next score (heterogeneity) gives us further insight into this issue of concentration by telling us what percentage of the passages come from the single most dominant novel. While the topic might be present in many different novels, it could be the case that most of the passages come from a single novel. Heterogeneity represents the percentage of passages that are not drawn from the single most dominant novel within the topic. The higher the heterogeneity score, then, the more balanced the topic is across different novels.[28] The average date (avg.date) gives us a sense of the temporal weight of the topic, while the standard deviation (sd.date) can help us see how broad the spread of dates are in the topic. A higher standard deviation suggests more temporal range for the topic.

The final two scores are designed to give us some sense of the semantic coherence of the topic. "Coherence" measures the co-document frequency over document frequency for the top twenty topic words associated with a topic. Higher negative numbers are associated, however counterintuitively, with more coherence. The more topic words appear together in documents as opposed to in their own documents, the more "coherent" the topic is thought to be (and the more it in theory correlates with expert opinion about the validity of the topics).[29] "Coherence" allows us to measure the interwovenness of the topic among documents. Finally, average similarity (avg.sim) measures the overall pairwise similarity between documents for which this topic is present above the stated threshold. The higher the score, the more similar documents are to one another according to their overall vocabulary distributions. This allows us to get a sense of the relatedness of documents beyond the topic words. Are these documents more or less semantically related beyond the topic vocabulary?

Given all these measures, we can see how Topic 150 is appealing for several reasons. While this topic is one of the smaller topics in terms of the overall number of words represented in the corpus (#tokens) or number of passages (#passages), it is among the highest in terms of spanning the most number of novels (#novels). Unlike many topics that are associated with a single work or a few from a given period, this topic is one of

Table 3.1 Summary of Topic 150 features

Feature	Score	Percentile
# tokens	203,348	14
# passages	30	34
# novels	19	93
heterogeneity	0.83	92
avg.date	1816	27
sd.date	36	91
coherence	−134.9	58
avg.similarity	0.735	10

the most diachronically rich. The average date of novels associated with the topic is 1816, suggesting it is concentrated in the late Romantic period. But it also has one of the highest standard deviations of date ranges (around 36 years). While anchored in the early nineteenth century, it moves across a broad range of historical time. Finally, it has slightly above average coherence in terms of the co-document frequency of the topic words (coherence), but scores very low in terms of the average similarity of the documents to each other. What this means is that the topic words occur relatively more often together within documents when compared to other topics—these are words that are somewhat traveling together—but the overall lexical composition of the documents tends to be more heterogeneous when compared to each other than for a majority of other topics. Alongside the topic's coherence (understood as the way the topic words appear more regularly together), there is an underlying semantic heterogeneity between the documents that is also interesting.

Where table 3.1 shows us the features that help describe the topic, table A.4 in the appendix shows us the documents with which this topic is most strongly associated. We can see, for example, how the topic moves from Goethe's *The Sorrows of Young Werther* to the world of the Romantic epistolary novel (Tieck's *William Lovell*, Hölderlin's *Hyperion*, Brentano's *Godwi*, and Bettina von Arnim's Goethe-homage, *Goethe's Briefwechsel mit einem Kind* [Goethe's Letters to a Child]), and into modernist incarnations like Rilke's *The Notebooks of Malte Laurids Brigge* and Ilse Frapan's novel of social protest, *Arbeit* [Work]. There is a generic coherence that is of interest here in terms of the epistolary novel—or its kin in the "notebook"—one that suggests an interesting insight into the way this type of novel has mattered to readers from the late eighteenth century to the early twentieth. While much of the scholarship on the epistolary novel concerns notions of an emerging publicity of the private or attention to modes of sentimental correspondence, this topic highlights the extent to which letters and notebooks serve as reflective media on some of the most elementary human categories, such as the distinction between life and death.

Topic 150 can be said to begin on October 27, 1774, in a letter from part two of the first edition of *The Sorrows of Young Werther*. It is the period in the novel that marks the beginning of Werther's end. In the second edition, Werther will explicitly remark a few letters earlier that he is now entering his own autumn. At the close of this letter, we encounter one of the more profound uses of natural imagery within the novel to convey a sense of individual collapse. In the readings that follow, I will be using a notation system to foreground the different categories to which the words in the passages belong: bold = words from Topic 150; underline = words that are distinctive of the documents using the technique of tf-idf;[30] and highlighting = words I have identified that appear to me to indicate a secondary conceptual concern. As I will argue, the explicit topic of life and death appears to bring with it an accompanying attention to a metaphorics of liquidity and aesthetic experience. One of my questions is how we can understand these "subtopics" or "hypertopics," that is, constellations of figures that are present within a topic but register at a different semantic level. Here is the passage in *Werther*:

When I go to the window and **look** upon the distant hills, the way the morning sun breaks through the mist from above and illuminates the **quiet** meadow, and the gentle stream winds its way toward me through the leafless pasture, oh when magnificent nature stands so lifelessly before me like a varnished painting, and all this joy cannot even pump a single drop of bliss into my brain, and this **man** stands there before **God**'s eyes like a dried-up spring, like an empty bucket! I have thrown myself so often on the ground and begged **God** for tears, like the farmer does for rain when heaven stands above him and around him the earth dies of thirst.[31]

The passage will end with Werther invoking Christ's famous intonation "My **God**, my **God**, why have you forsaken me?" By December he will be dead.

This passage is about the loss of something graspable, in the sense of that which mediates between us and our beliefs, between ourselves and something greater than ourselves. The loss of Lotte, the knowledge that Werther cannot have her, marks the initiation of this crisis of belief. "Only **God** knows how someone can do it, to see so much grace circling around oneself and not be able to touch it [*nicht zugreifen zu dürfen*]. And yet is not the act of grasping the most *natural* drive of mankind?"

The loss of mediation, of the graspable, will be expressed in Werther's words as a crisis of liquidity, as a loss of circulation—"this man stands there before God's eyes like a dried-up spring." Something is no longer able to pass through, to circulate, because it lacks a real correspondent. Nature is no longer a hypothetical, a point of reverie, as it had been in the novel's first half, but a "varnished image," now lifeless. Such mortification will later be compared at the end of the passage with the bitter cup served by God to his son: "Was not the cup too bitter for **God** when it touched his **human** lips? Why should I deceive myself and pretend as though it tastes sweet to me?" The spring at the opening of the passage, the eternally renewable source, has turned by the passage's conclusion to the poisonous draft—to the irreversible, linear, and finite movement that marks out the end of one's being.

Werther will refer to this new state as one of "trembling between Being and Not-Being," where "the past, like a lightning bolt, illuminates the gloomy abyss of the future and everything sinks around me as the **world** expires." The letter is the medium that articulates this sentiment of death in life, of becoming the very mediation that is no longer at hand. And yet Werther cannot live in this moment of betweenness; he cannot be the letter. The loss of Lotte's graspability, and the failures of epistolary circulation, mark the end of Werther.

It is precisely this experience of the in-between, of experiencing the negation of being as something livable, that will move from something destructive in *Werther* to a positive end state of the Romantic novel. How to dwell in this middle, to occupy the center of the life/death topos, will emerge as a foundational Romantic question. In a passage from Hölderlin's *Hyperion*, the novel most strongly represented in this topic (see table A.4), Diotima tells Hyperion: "Don't you see now, how poor and how rich you are? why you must be proud and despondent? why *joy* and sorrow alternate so terribly within you? It's because you have everything and nothing, because the phantom of the golden age that is supposed to come belongs to you and yet it is not there."[32] It is precisely this dual state, of everything and nothing, life and death, that represents Hyperion's being. "You are a citizen in the regions of justice and beauty," says Diotima, "a **God** among **Gods** in the beautiful *dreams* that elude you at daybreak." The aim of Hölderlin's novel is to find a language to articulate how one can dwell in this topos of the in-between.

In Ludwig Tieck's epistolary novel *William Lovell* (1796), we see how the main character experiences a similar state of liminal nothingness, once again using the metaphor of the spring, only this time as a magic mirror: "Yes, Eduard, do not poke fun of my weaknesses. In these moments I am superstitious like a child. **Night** and *loneliness* have strained my fantasy, I peer like a medium down into the **deep** spring of the future, I see forms that rise up to me, friendly and earnest, but also an entire army of terrible images. The straight line of my **life** begins to entwine itself in knots, and to undo them I may sacrifice my existence in vain."[33] And in a passage from Brentano's *Godwi* (1802), the hermit, Sent, will sing of life after death using similar language of liquidity (waves and drowning) to articulate this dream of dying in art.[34] Far from the drying out that marked Werther's fall, here the liquid serves as a form of being consumed by some externality, as a form of life in death. The excess of fluidity and the totality of fluidity—*becoming fluid*—serve as the conditions of Romantic insight.[35]

As the epistolary novel develops over the course of several decades from *Werther* onward, the sense of self-loss that was essential to Goethe's novel will be rewritten in its Romantic aftermath as a foundational aspect of aesthetic experience. It is a shift that, according to our topic space, appears to be articulated through the language of liquidity, as the dried-out spring turns to the consumptive drowning of art's immersiveness. While such a vocabulary does not appear in either the leading topic words or the most distinctive words, it remains submerged, so to speak, within the topic space as a recurring metaphorical constellation. The very fluidity of the topic as topic is arguably what allows for these heterogeneous configurations to exist alongside or within one another.

To push the point still further, we can look at another passage from Hölderlin's *Hyperion* associated with this topic.[36] Here we read about the young man's desire to preserve aesthetic experience, once more invoking the metaphorics of fluidity. "I went into the forest along the trickling water where it dribbled over cliffs and slid harmlessly over the pebbles . . . Here—I wish I could speak, my dear Bellarmin! I wish I could calmly write to you!" A few sentences later, Hyperion will summarize what this moment has meant to him: "I have *sacredly* preserved it! like a Palladium, I have carried it around with me, this *divinity* that appeared to me! And from now on when fate grasps me and hurls me from one abyss to another and all my strength and thoughts are drowned, then this *singular* thing shall survive beyond me within me, and shine within me and reign in **eternal**, indestructible clarity!"[37] Later, Hyperion will call this the "peace of beauty." "Do you know his name? the name of that which is one and everything? His name is beauty."

Here we see Hölderlin giving voice to the paradox of aesthetic loss as a form of plenitude in the most compressed terms possible: "this singular thing shall survive beyond me within me [*so soll dies Einzige doch mich selber überleben in mir*]." For the Romantic novel, we die into art. In doing so, something persists, both ourselves and something more than ourselves. So can a single thing [*dies Einzige*] be multiple, art an entry into plenitude.

One of the more interesting results of this analysis is that if we cluster the documents within this topic by their topic words, we see a unique cluster emerge that consists of the initial passage from *Werther* I cited above, the first one from *Hyperion* that was discussed, and one from Rilke's *Malte Laurids Brigge*.[38] Ulrich Fülleborn was the first to draw this connection between these three novels, one that has largely lain dormant in the scholarship, and the topic model offers us a way, not only of "rediscovering" this connection, but of understanding one of the particular ways that these novels intersect.[39] In the passage from Rilke's *Malte* that is included in this topic, we encounter the figure of the *mouleur*, where we hear of a plaster cast of a drowned woman that leads to an extended reflection on the self-contained mind and the possibility of pure art. The facial cast of the women fatally submerged by water becomes the object of reverie to imagine an ideal form of music, one that, the narrator tells us, precipitates from the heavens. "**World**-consummator: as that which comes down as rain over the **earth** and upon the waters, falling carelessly, at random,—inevitably rises again, invisible and joyous, out of all things, and ascends and floats and forms the **heavens**: so our precipitations rose out of you, and vaulted the **world** with music."[40] Once again we can see the metaphorics of liquidity at work here, as the notion of consummation emerges out of the static plaster cast of the submerged woman. It is this music, "your music," Malte says, that stands in distinction to modern concertgoers, "whose sterile hearing fornicates and **never** conceives, as the semen spurts out onto them and they lie beneath it like whores, playing with it." This is fluidity (and art) of the wrong kind.

Throughout this topic, in other words, we can pick up on this, we might say, uncanny association between art and liquidity. Topic models are expressly concerned with identifying "latent" associations in language, and it is decidedly interesting that we can find other latent constellations within the topic's own latent associations. This is not to say that all documents in the topic are concerned with fluidity, or that there are not other concerns circulating within this topical space. For example, a sense of "departure" looms large in passages by Jean Paul as well as Hölderlin, and "corporality" seems interestingly bound up with questions of life and spirit. In Jean Paul's *Hesperus*, for example, there is a madman named "Totengebein" who believes death is following him and wants to grab his left hand, and in Ilse Frapan's *Work*, her heroine, an obstetrician, moves over the course of a single passage from thoughts of taking her life with her own hand to facilitating a birth of a near-dead mother, where the transition from death to life is figured through the removal of her gloves [*Handschuhe*]: "But as she removed her hat and gloves, she had an entirely different face. The excitement is as though washed away. Here there is only deep seriousness and an entering into her task." In *Work*, the bearing of one's hands becomes the means of entry into the primordial maintenance of life's cyclicality.

Liquidity and the aesthetic are thus not the only concerns within this topic, but they do seem to persist with a surprising degree of regularity. Amid the vocabulary of life, death, eternity, man, God, and the soul, a concern with the immanent transcendence of art will gradually emerge. Although the word *beauty* is not contained in the leading words of this topic (it ranks 66th), it becomes in many ways implicitly associated with this configuration, one that, as we have seen, is repeatedly manifested through the language of liquidity. The retreat of the graspable medium in *Werther*—of art as a container of liquid, of art as jug or cup—is replaced in the Romantic and post-Romantic novel by a sense of liquidity as an immersive, transformative totality—art as a form of drowning.[41] As Gaston Bachelard writes in his *Water and Dreams*: "A being dedicated to water is a being in flux. He dies every minute."[42] This is one of the longings of language in the Romantic epistolary novel, made visible by the topos of life and death.[43]

Topical Dispersion

In the space remaining, I want to continue this exploration of the conceptual diversity of the computational topic, but I will be doing so using more computationally driven methods. Here again the practice of close reading, which was framed by the distant model, informs the process of further model-building. Instead of focusing on within-topic diversity, as I have just done, I am interested in better understanding cross-topic relationships, the ways in which topics intersect, disperse, and "take shape" out of other topics. How can we begin to think of topoi not as fixed, immutable spaces, but as mutable arrangements, as "likelihoods," in a more literal, place-bound sense?

The first approach I will use is to build a network of the relationships between topics (fig. 3.1).[44] In this approach, I use what is called a bipartite graph, where nodes are drawn from two distinct types of groups. In this case, nodes can be either topics or documents. An edge is then

CHAPTER THREE

[Network diagram with labeled nodes:]

- innocence
- minister, child, old
- heart, spirit, love
- all, always, love
- all, love, heart
- heart, soul, man
- life, eternal, death
- man, beauty
- all, being, poetry
- spirit, ach, beauty
- man, soul, spirit
- love, beauty, life
- travel
- hand, called, arm
- man, law

3.1 Network diagram of topic-to-document relationships in 14,888 passages from German novels. Only topics that have a connection with Topic 150 are depicted.

drawn between the two types of nodes when a topic is strongly present in a document (defined according to our threshold as being more than 20% likely to be in that document). Topics are thus connected through documents in which two or more topics are strongly co-present. The graph presented here does not show the entire model, but contains only topics that are associated with documents associated with Topic 150. This is the most immediate (1-hop) topical neighborhood of Topic 150. The labels used for the topics are drawn from the leading words in the topics and are sized according to the number of documents associated with those topics (bigger means more documents contain this topic).

Looking more closely at the graph, we can see how the topic of life and death resides within a larger constellation of concerns that are more explicitly associated with beauty, as well as notions of *man*, *spirit*, and virtue (*innocence*) [*Mensch, Geist, Unschuld*]. The more latent concerns with beauty that I saw in Topic 150 become far more explicitly associated with neighboring topics, which are also more prominent overall. Aesthetics, ethics, and the (spiritual or intellectual) knowledge of man are tightly bound together in this emerging anthropological field of the epistolary novel. The acknowledgment of human finitude stands at its center. In this, we can see a profound confirmation of Michel Foucault's insight that the human sciences emerge in the nineteenth century from knowledge of human time. As Foucault writes, "Man's finitude is heralded—imperiously so—in the positivity of knowledge."[45] Man and beauty, anthropology and aesthetics, are born, according to this graph of the novel's topical concerns, around the knowledge of life and death.

If this is the larger topological field in which the knowledge of life and death lives in the novel, I am also interested in understanding the semantic coherence or constellations that underlie the topics themselves. Rather than treat topics as coherent nodes—dimensionless in their unity—we can also visualize the relations between the units of the individual topics. How do the topic words relate to each other, not simply in their dominant or most likely state, but also when they disperse and become weakly present? How can we observe topics assembling and falling apart, tracing the boundaries of spatial and semantic coherence?

Topic models are useful in that they provide a list of words with associated probabilities of being present in a topic. They help us see a hierarchical relationship between words (this word is more likely than that word). But they do not tell us much about the individual associations between words, the configurations of their occurrences that contribute to the topic's larger meaning. To do so, we can once again use a vector space model to translate the probabilistic list of words into a spatial representation of the correlations between words.[46] In this way, we can begin to get a better sense of the topic as a larger semantic field, that is, as a topos.

In figure 3.2, we see a representation of the correlations between the leading topic words in the thirty documents where this topic is most strongly present (more than 20% likely) using the process of multidimensional scaling that was discussed in the previous chapter. The closer two words are to each other, the more similar they are to the words around them in terms of their quantitative occurrences across these documents (the more they "correlate"). It is important to point out that in representing the terms in this way, we lose some local information about individual relationships between words, but we do so in order to gain more global information about the overall relationships between them. The black/gray distinction represents two clusters that were detected using the method of hierarchical clustering, suggesting that if we had to partition these words into two groups, this is one way to do so. What we

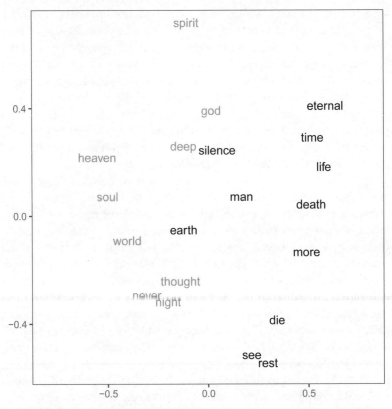

3.2 Correlation of words associated with Topic 130 represented using multidimensional scaling. Black/gray represents predicted clusters.

see happening when we do so are the more spiritual terms, such as *soul, heaven, god*, and *spirit*, occupying the left portion of the graph, while the more temporally oriented terms like *time, eternal, life,* and *death* occupy the right. Interestingly, in this model, "man" maintains something of a middle ground between these two worlds. The pairings that are generated are also of interest in their own right: *night & thought, rest & sight, silence & depth,* and of course *life & death.*

In *Semiotics and the Philosophy of Language*, Umberto Eco offers a thoughtful critique of the history of the Porphyrian language tree—the tradition in which we think of the meaning of words hierarchically as a series of descending branches. This is still a strong component of the way linguists continue to think about language (and one, for example, underlying the hypernymic structure of WordNet used in the first chapter).

For Eco, such hierarchies nevertheless can never truly account for language's diversity. As he writes, "There is no bidimensional tree able to represent the global semantic competence of a given culture."[47] Instead, Eco reminds us, we are left only with manifestations, contingent constructions that could always be otherwise. As he writes, paraphrasing Thomas Aquinas, "essential differences cannot be known . . . We can only know them through the effects (accidents) that they produce" (69). In the multidimensional topical space seen above, words do not belong to fixed branches of a bifurcating tree; rather, they exist in a more horizontal and shifting relationship of complementarity. While these spaces cannot tell us everything we want to know about these words, and while much information is lost in the reduction from twenty to two dimensions, they can still give us a feeling for the semantic associations of the topical space *as a space*. There are no strict connections or dependencies here, just proximities and distances and configurations.

As we will see, such configurations are also not permanent, but can mutate depending on the nature of the documents being queried. Rather than look at the topical field in its strongest state, we can also observe what the semantic arrangement looks like when we examine documents for which this topic is only weakly present (defined as a likelihood between 10 and 12%) (fig. 3.3).[48] Here we see how the clustering that was at work in the stronger state disperses, with the central pairing of *life* and *death* that defined that earlier state also undone (they now occupy opposite poles). *Man* too has moved from its central, mediating position in relation to all the other terms outward toward a semantic world marked by a strong binary between earth and heaven, time and God.

Looking at the novels that are weakly associated with Topic 150, one of the epistolary novels that appears here that was not in the topic in its strong state is Sophie Mereau's *Amanda und Eduard* (1803). Women writers were only sparsely represented in Topic 150, and *Amanda und Eduard*'s presence in its weaker state suggests alternative narratives that might be surrounding epistolary novels, especially those written by women (or even about women, as in the case of Fr. Schlegel's *Lucinde*, also present here).[49] Topic 73 is one of the most common topics within Mereau's novel, appearing in both this liminal range and well above the threshold. *Amanda und Eduard*, or at least certain portions of it, appears to act as a kind of bridge between these topics in their unformed states. It points to the ways these two topics have something to do with each other—not in their distinctive co-presence with each other, as we saw in the network diagram where topics are joined by documents in which they

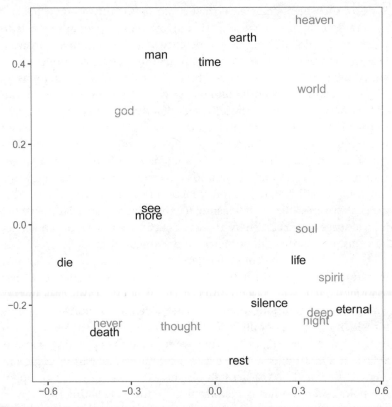

3.3 Correlation of Topic 150 words in documents where the topic is only weakly present.

are present in their most likely states. Rather, it suggests that these topics overlap in more diffuse ways—they are "related," but not in a strict genealogical or *linked* sense. They share porous boundaries.

If we graph these two topics together in documents in which they are *weakly* present (again defined as a 10–12% probability of being present), we see how the topics are still largely separate even in their weak state (fig. 3.4). The words that do appear to cross over, such as *soul* and *deep*, are, interestingly, those words that are present in both topics. We can see how they begin to acquire new meaning as they become attached to the emotional states of Topic 73, such as *feeling* and *love*, and move away from the theological concepts with which they were associated in Topic 150, such as *heaven* and *God* (fig. 3.2). *Death*, too, loses its antinomial correlations with *life*, *time*, and *more*, and assumes a more semantically familiar cluster of human finitude (die, death, man, earth). These words are becoming more literal, while the soul is becoming interiorized and less theological.

Finally, we can look at how Topic 73 appears in its strong state and compare it to this earlier state of emergence (fig. 3.5). Here we see how the topic cleanly splits into two distinct clusters, with the group in the upper right focusing on an emotionally laden visual beauty. *Beauty* and *Bild* (image) are now strongly correlated along with a few emotional modifiers, such as charming [*lieblich*] and allure [*reiz*], along with *feeling* itself. *Love* and *heart* are brought closely together in a kind of epochal Romantic coupling, while the *soul* is now more strongly associated with the *act* of looking, one that focuses more on interiority rather than appearances (where *appear* or *schien* appears in the other cluster). Surface and depth combine as two distinct, interactive visual planes in this topos. The binary hierarchy between earth and heaven that accompanied

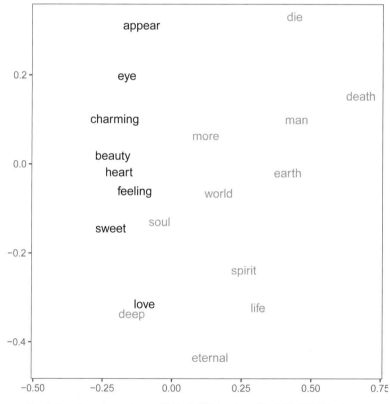

3.4   Correlation of words associated with Topic 150 (gray) and Topic 73 (black) in documents where the topics are only weakly present.

CHAPTER THREE

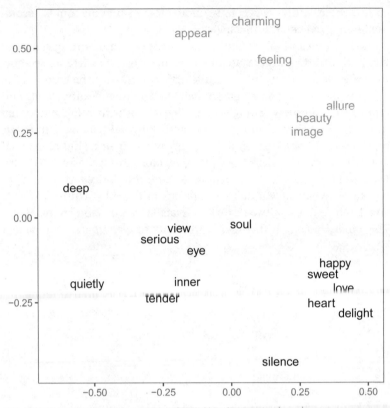

3.3 Correlation of words associated with Topic 73 in documents where the topic is strongly present. Colors represent predicted clusters.

Topic 150's concerns with life and death is translated into the more interpersonal concerns with inner and outer forms of aesthetic experience.

By way of conclusion, I want to look more closely at the passage from *Amanda und Eduard* that is most strongly associated with Topic 73 but where Topic 150 is also still weakly present. What happens when concerns of life and death disperse, but do not altogether disappear, and those of beauty and the soul emerge in their stead? What does this relationality of different topical spaces look like on the page?

The passage consists of two parts, including the closing moments of a letter from Eduard to his friend Barton and the opening moments of a letter from Amanda to Julie. In many ways they parallel each other. Eduard describes an encounter he has while riding alone in the country and seeing a carriage pass by in which he observes a beautiful sleeping woman and her older companion. It represents the quintessential moment of the male gaze (here my notation uses bold for Topic 73 words, italics for Topic 150 words, and gray for liquidity words): "I turned around, the carriage drove slowly ahead, and I had *time* to *calmly* observe her. Sleep poured a **charming** indeterminacy [*Unbestimmtheit*] over her **beautiful** features, where no dominant expression was visible. It occurred to me that a truly **beautiful** face should *never* have any other expression than one of pure harmony received from nature."⁵⁰ As the carriage moves, time slows down, allowing Eduard to observe the female visage in a state of rest, in this in-between state between life and death, which he sees as one of pure and natural harmony. It is a world of appearances, beauty and *Bild* perfectly conjoined.

In her subsequent letter, Amanda will describe her desire to no longer be alone in the world: "And I am alone! alone in this most *vital* nature, whose happiest **images** are covered over by this **feeling**, as if with a melancholy veil. These fragrant bowers desire **affectionate** conversation, these **alluring** pathways want for meaning.—Alas, perhaps only a single hedge in bloom or a single pathway separates me from the object that would be worthy of sharing my **feelings**. Perhaps he too wanders alone, with that **beautiful**, unsatisfied **heart**, astonished to find *dead* nature so *alive* and the *vital world* so dead!" The mixture of life and death now occurs through the imaginary encounter with another, an encounter that lies at the core of the sentimental vocabulary of beauty and the heart. Indeed, it is the encounter, or *Begegnung*, that unifies both of these experiences, one actual, the other imaginary. That which stands opposite to us, the object or *"Gegenstand,"* in Amanda's words, is what lends meaning to one's surroundings. "The pleasant emotion [*Rührung*], which moved my heart," writes Eduard, "poured a higher charm upon all of nature surrounding me." In addition to yet another gesture toward liquidity here (in the pouring out of charm upon the landscape), we see how it is a vocabulary of contact, *Rührung*, or the act of agitation, that is at stake in the reflection on beauty. No longer the dried-up wellspring of Wertherian isolation or the drowning self-loss of the male Romantic epistolary novel, in Sophie Mereau's rendering it is *opposition*, that which is external to ourselves, that serves as the source of aesthetic insight. This is what activates the vocabulary of observation and appearance that was present in, but not central to, the topic of life and death. Life in the female epistolary novel is enlivened through the notion of the *encounter*, with both the beautiful but also the unknown. Immersion gives way to the metaphorics of agitation, to experiences marked by contact with an exteriority, the strangeness of what it means to be a woman in a highly disciplined and overseen world. Agitation and encounter mark out one's finite place in the world. Beauty is that which makes access to opposition possible.

Conclusion

This chapter has tried to show what happens when we interpretively inhabit the spaces of computational topics. In a particular sense, I have engaged with the topos of life and death that critics since Foucault have routinely identified as one of the core themes or "transcendentals" of modern life. According to this topic's activation in the epistolary novel of the nineteenth century, we see how on the one hand it largely confirms Foucault's hypothesis about an emerging discourse of depth and a new verticality of culture. The "immersiveness" of aesthetic experience, its fundamental liquidity, strongly aligns with the surface/depth model of reading that becomes normative during this period. Such immersion would paradoxically serve as a medium of self-dissolution in the Romantic period, of achieving a union with some higher spiritual state. This is contemporary criticism's Romantic legacy.

But when we observe the intersections of topics in their more disperse states, we can also see *other* kinds of narrative models taking shape during this period, ones that are also highly gendered. Instead of conceptualizing a discourse on human being through the terms of either finitude or spirit, for the female epistolary novel it is the notion of "encounter" or *Begegnung*, the state of being opposite to someone, that provides the grounds of aesthetic insight. Agitation and otherness—a horizontal being opposed to—replace a sense of immersion and consumption as art's entry point into human understanding. The tactile and the object, the possibility of being touched but also estranged, replace the synesthetic communal longings of male liquidity. This is another potentiality that resides within the nineteenth-century topos of life and death.

For Curtius, the *Topik* was something that transcended both time and space. It was a form of cultural stability that resisted both historical change or individual whim. The computational topic appears to situate us far from this world of "immutable mobiles" (to borrow Latour's phrase). These are not stable lexical *things*. Instead, the topical spaces of computational topic modeling allow us to intuit a greater sense of the interwovenness of topical space, the differences that reside within these zones of similarity and the similarities that transcend the distinctions we traditionally draw around authors, genres, periods, books, or even "ideas." Computational topics reframe the post-structural project of intertexuality, not so much as the presence of one text in another, as a form of citation, but in a sense of dedifferentiation, where the text contains a multiplicity within itself that can never entirely be accounted for. Quantification serves in this sense not as an entry into the empirical and the definitive, but the conjectural and the interpretive.

My hope is that when we begin to see reading in this way, indeed, when we see reading as a conjunction of seeing and reading, of space and text, the usefulness of computational topics will appear to reside less in their ability to retrieve literary "information," and more in how they reveal a criticism of uncommonness. Meaning within topological fields is a matter of both configuration and dispersion, the process through which ideas do not entirely contain themselves and thereby produce difference. The more we are able to understand language at this level, the more we will be able to attend to the repetitions and their differences that underpin literature. Literature, seen in this way, is an art of the centrifugal.

FOUR

Fictionality (Sense)

"There is no textual property, syntactical or semantic, that will identify a text as a work of fiction." JOHN SEARLE

The distinction between fiction and nonfiction, between a text that is true and one that is not, is one of the oldest on record. Ever since we started thinking about the act of narration, we have addressed the related meanings of truth and imagination. This is what Aristotle designated as the difference between the communicative use of language (*legein*) and its creative use (*poiein*).[1] For millennia, we have been debating whether there are inherent features of being fictional or whether it is simply a matter of intention, that perhaps there is nothing unique to the language of fictional discourse after all. How do we know when a text is signaling that it is "true" or, by extension, "not true"? And what might quantity have to tell us about this most elementary of distinctions?

Consider for example the following two passages:

A
On the short ferry ride from Buckley Bay to Denman Island, Juliet got out of her car and stood at the front of the boat, in the summer breeze. A woman standing there recognized her, and they began to talk. It is not unusual for people to take a second look at Juliet and wonder where they've seen her before, and sometimes, to remember.

B
Jeff is 24, tall and fit, with shaggy brown hair and an easy smile. After graduating from Brown three years ago, with an honors degree in his-

tory and anthropology, he moved back home to the Boston suburbs and started looking for a job. After several months, he found one, as a sales representative for a small Internet provider. He stays in touch with friends from college by text message and email, and still heads downtown on weekends to hang out at Boston's "Brown bars." "It's kinda like I never left college," he says, with a mixture of resignation and pleasure. "Same friends, same aimlessness."

At first glance, these passages share a good deal in common. Both use single proper names (Jeff / Juliet) and markers of place (Boston / Denman Island). They each use a number of pronouns (her, they / he), an occasional adjective (short ferry ride, summer breeze / shaggy brown hair, easy smile), as well as the past and present tense (stood, began, is / is, started, moved). While the second passage uses dialogue, it is not unreasonable to assume that the first text might at some point, too. And both passages seem to offer some kind of psychological underpinning to the description, whether it is Jeff's malaise as a "sales rep" or the woman who recognizes Juliet in the first passage. Both make claims on our thinking about people and personality.

And yet few readers would have difficulty guessing that passage B is from a work of nonfiction (Michael Kimmel's *Guyland*) and passage A from a work of fiction (Alice Munro's "Silence"). What is it, then, that makes this so obvious?

There are many reasons we might think of to explain this difference. Table 4.1 gives us an abbreviated list of some possible features we might use.[2] While this is just a start, it already begins to tell us something about the distinctive linguistic and syntactic qualities of these passages. Munro's text now looks more verb-ish than Kimmel's ("verb"), with four times as many present-tense verbs as Kimmel ("present"). Her sentences are also somewhat longer ("wps"), and more pronoun heavy ("pronoun"). She uses a more distinctive vocabulary of cognitive insight ("recognize," "remember," "wonder"), but also sociability ("woman," "people," but also "talk"). Kimmel, on the other hand, uses many more six-letter words, slightly more commas and periods, more numbers, and a greater vocabulary of affect and work. We might think the literary text would be more affective, but part of Munro's art is submerging feelings so that they are more implicit than explicit ("and wonder where they've seen her before, and sometimes, to remember").

This chapter is about trying to understand these differences at larger scale. Rather than look at a single example, as I have done above, or even several of them, I will be using a collection of roughly 28,000 documents, both fictional and nonfictional, to better understand what distinguishes

Table 4.1 Selected features comparing a passage from Michael Kimmel's nonfictional *Guyland* to Alice Munro's "Silence." Values represent the percentage of words in the passage that belong to a feature.

Feature	Kimmel (%)	Munro (%)	Difference (%)
verb	4.8	14.2	9.4
present	1.9	7.8	5.9
wps	17.3	21.3	4.0
social	8.7	12.5	3.8
insight	0.9	4.7	3.8
pronoun	5.8	9.4	3.6
work	3.9	0	−3.9
time	11.5	6.3	−5.2
posemo	5.8	0	−5.8
affect	6.7	0	−6.7
sixletter	21.2	10.9	−10.3

fictional writing from its nonfictional counterpart. What role does quantity play when writers signal to readers that a story is true or not true (which is different from the actual truth content of a story)? Much of my emphasis will focus on the novel as one of the dominant forms of fictional writing from the nineteenth century to the present. Beginning around 1800, the point at which we know the novel begins its inexorable quantitative rise to prominence, what makes it unique as a form of fictional discourse?[3]

Questions about the nature of fictional speech reached something of a high point in the 1970s and early 1980s, with numerous works in the philosophy of language reflecting on the linguistic cues that marked out a text's truth claims.[4] At stake in this endeavor was an attempt to define and thus potentially control for the reliability of language, the ability to distinguish between the truthful and untruthful content of speech. The work of John Searle became a landmark within this movement, providing a framework that was deeply indebted to the theory of speech acts inherited from J. L. Austin.

For Searle and the community of philosophers gathered around him, the differences between fictional and nonfictional discourse did not depend on the actual content of the speech. Instead, it depended on the combined intentionality of the speaker and receiver, what were known as illocutionary and perlocutionary acts. (We might think of these as frameworks for producing and receiving speech.) As Searle writes, "The utterance acts in fiction are indistinguishable from the utterance acts of serious discourse, and it is for that reason that there is no textual property that will identify a stretch of discourse as a work of fiction."[5] For the philosophers of language, fictionality was not a distinct use of language, but depended on the intentions of both writers and readers and the way those intentions were communicated beyond the boundaries of a text. Their goal was to show how the same set of words could mean very different things depending on the speaker's intentions.

For literary theorists of about the same time, literature, as a subset of fictional discourse, similarly came to be defined as an indistinguishable textual entity from the larger category of "writing." Searle's and Austin's speech-act theory was used to generate a more general critique of literary essentialism, that there were unique and potentially timeless qualities to works of literature. As Jacques Derrida would write, explicitly invoking Searle's philosophy, "No exposition, no discursive form is intrinsically or essentially literary before and outside of the function it is assigned or recognized by a right, that is, a specific intentionality inscribed directly on the social body." Derrida would then continue, "This is the hypothesis I would like to test and submit to your discussion. There is no essence or substance of literature: literature is not. It does not exist."[6] For Derrida and much of the post-structural criticism that followed, literature was the product not of a definable set of features, but a social set of intentions, the frameworks of production and reception that underpinned Searle's speech acts.[7] Translating Searle's position on discursive statements into literary interpretation more generally, literature was seen as liberatory precisely because it was irreducible to any kind of pattern, habit, or idiolect.[8]

This chapter will make the exact opposite claim, one that will assume three basic parts. First, from the perspective of machine learning, fictionality is a *highly* legible category at the level of linguistic content ("lexis," in Aristotelian terminology). When we take into account a sufficient number of words, we can build predictive models that can identify works of fiction with greater than 95% accuracy (mirroring in many ways the ease with which readers are able to do the same, as in my initial experiment above).[9] The ambiguity that appears to exist at the level of the sentence or "utterance" (what Searle rather vaguely called a "stretch of discourse") no longer holds when we observe writing at a different level of scale. This is one of those classic instances that Stanley Cavell identified where the philosopher's example is simply too simple to explain something in the real world. Given enough words, the intentionality that is supposed to reside beyond the semantic content of a statement is indeed largely recoverable from that semantic content. How many words is enough? As figure 4.1 shows, already by 100 words a machine can tell the

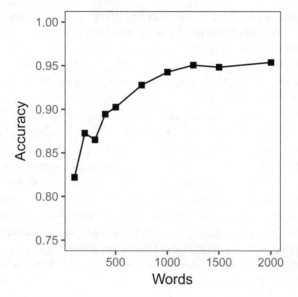

4.1 Accuracy of predicting fictional texts using an increasing number of words from the beginning of a document.

difference between fiction and nonfiction with over 82% accuracy. Its maximum performance is achieved around 1,250 words, or roughly 3–4 pages of text.[10]

This is not to say that the actual truth content of writing is knowable through its linguistic features (one could, for example, simply change the dates of a work of history for it to no longer be true, but still be understood as a work of nonfiction). But we can recover the *intended* nature of a document's truth claims with a great deal of accuracy. While in theory literature is an open-ended system, where any piece of writing can potentially be included, in practice there is a tremendous coherence and predictability, a determinacy, to the nature of literary narrative, even the most canonical of writings. When writers wish to signal to readers that a work is true or not true, they do so with remarkable regularity.

Second, not only is such legibility coherent across different kinds of narrative and, as we will see, languages, it also appears to have been surprisingly stable for at least two hundred years. There is a degree of *immutability* to fictional discourse that is also worth taking note of. This is not to say that there have not been stylistic changes across time among and between works of fiction, changes to which I will return at the close of the chapter. But the distinctiveness of fiction itself has remained significantly stable over the two-century window I am looking at here. If we train learning algorithms on nineteenth-century texts, for example, we can still recognize contemporary novels with an impressive degree of accuracy (about 90%), even if that performance does decrease (history still matters). Indeed, the very features that seem to indicate the uniqueness of fictionality in the nineteenth century appear to be either increasing over time or largely holding steady. Such continuity has important and still largely unaddressed implications for how we think about both genre and literary periodization.[11]

Finally, understanding the particular nature of this legibility—*how* fiction distinguishes itself as a cultural practice—has important implications for our scholarly narratives. As we will see, fiction's stability, and the novel's in particular, appears to be based on what I will be calling a sense of "phenomenological investment." Seen quantitatively, the particular nature of fictional discourse since the nineteenth century has been its profound investment not simply with the world around us, but with our perceptual encounter with that world, the way "making sense" is explicitly related with the physical senses. As Dorrit Cohn would write two decades ago, "Fiction is recognizable as fiction only if and when it actualizes its focalizing potential."[12] When we look more closely at what distinguishes the novel from other kinds of fictional discourse in its classical nineteenth-century phase—what allows the novel to stand out during its quantitative rise to prominence—we see a secondary investment in a related notion of prevarication—a doubting, testing, hypothetical relationship to the world.

To think about fictionality and the novel in these terms puts pressure on some of the more common scholarly refrains of the recent past. Long-standing questions about the novel's "realism," that is, the extent to which and the means through which a novel reproduces a given environment, give way in this view to the novel's concerns with a "dramatization of encounter"—both with others and with the world, indeed, with otherness more broadly speaking. Rather than think of the novel as a genre based on its relationship to a knowledge of things, thing-theory serving here as a translation of the rise of realism into new terms, according to its quantitatively meaningful components the novel appears far more self-referential in nature, offering us access to the knowledge of knowing. It renders explicit an experience of the hypothetical, a testing as well as sensing relationship to the world.[13] While this may have traditionally been how we thought about a small subset of modernist experiments, it is significant that this insight holds even across the most canonical "realist" novels of the nineteenth century.

This chapter thus suggests that certain canonical positions within the history of novel scholarship need to be rethought or at least subject to revision in light of an emerging computational understanding of the novel. Whether it is Catherine Gallagher's argument about the novel's ambiguous relationship to its own fictionality; the post-structural investment in literature's negativity, as when critics speak of the novel as the "genre of no genre"; thing-theory's emphasis on what Elaine Freedgood has called the "denotative, literal, and technical" language of the novel; or Ian Watt's still-influential position on the novel's referentiality, as when he writes, "It would appear then that the function of language is much more largely referential in the novel than in other literary forms," computation presents a very different portrait of the novel's importance as a type of fictional discourse.[14]

The point is not that these positions are unfounded—it is certain that for some novels these ways of being may indeed be predominant, just as it is certain that for many novels these ways of being may be operative some of the time. But if we try to understand what makes the novel stand out from other types of ostensibly true writing or even other types of fictional texts—if we try to generalize about the novel as a genre—then at least since the turn of the nineteenth century we are seeing something altogether different at stake. According to the samples and models used here, the novel's mattering is not primarily grounded in its positive representation of the world, that is, in its mimetic utility, its ability to simulate something (as in seventeenth-century debates about *vraisemblance*). Nor is it grounded in a kind of post-structural negativity—the novel is unrecognizable as a distinct and stable category, a reflection of literature's more general negative capability. Rather, the novel can best be described through its investment in the negation of the certainty of its own worldliness. It is grounded in an appeal to an *embodied* encounter rather than reality itself. In doing so, it is precisely the referentiality of language that is being bracketed in the novel, not ambiguously, but programmatically, even in novels that are widely considered to be the most realistic.

Prediction and Description

This, then, is the threefold argument I would like to make in this chapter, which I will label the coherence hypothesis, the immutability hypothesis, and the phenomenological hypothesis. I label them hypotheses because whatever proof I offer here cannot possibly exhaust their validity. But the hypothesis is a useful framework in which to establish an interpretive framework, what Merleau-Ponty called the "situation of perception," the ability to occupy a shared observational space ("to stand where I am," in Merleau-Ponty's words).[15] In order to test these beliefs—and I am well aware of the circularity at work here to test a genre whose essence is a testing relationship to the world—I will be using a combination of both predictive and descriptive methods. As Ted Underwood has shown, predictive models, such as those employed in the process of machine learning, are important because they allow us to engage in the process of classification, of what it means to define a group of texts as a coherent entity and to understand the degree of coherence that surrounds that group according to certain predefined conditions.[16] Predictive models allow us to say with how much certainty we can identify texts that belong to a specific group and under what criteria. The more certainty there is, the more cohesive the category is thought to be. Classification is an important dimension of literary study because it allows us to understand the degree of coherence behind our conceptual categories.

Descriptive models, on the other hand, are useful because they allow us to qualify distinctive features of one group when compared with another without engaging in the act of classification. They can tell us which features are distinctive of one group versus another, but they do not do so in order to make claims about the overall uniqueness of that group. Instead of defining a text or group of texts—the novel is X or the novel is this predictable—these qualities help describe the behavior of a group according to more individualized criteria. This too is valuable because it allows us to understand the components that make one group different from another but that do not necessarily lead to categorical differences. Explaining predictions—how a computer arrived at an estimate about which class a text belongs to—is quite challenging. Explaining individual differences is far more straightforward. It is their combination, I would argue, that allows us to think both categorically—about the relative coherence of writing under certain conditions—as well as qualitatively about the specific aspects of writing regardless of drawing definitive boundaries around things. Description is in many ways much closer to the traditional task of literary criticism, but prediction can and should be a useful tool in the literary analyst's tool kit.

The data that I will be using for this chapter has been selected to understand the nature of fictionality across different types of writing. It is summarized in table A.5 in the appendix. The aim is to see if the results here hold across time, different languages, and different sample sizes. Overall, the data consists of about 28,000 documents, dating from the late eighteenth century to the early twenty-first, written in both English and German. The collections contain different kinds of fictional and nonfictional writing, including novels, histories, philosophy, advice books, novellas, fairy tales, and classical epics translated into prose, among other kinds of writing (though not including encyclopedias or cookbooks). Together, the texts can be grouped into four principal categories.

The first collection represents a canonical set of nineteenth-century writing of about 600 documents in both German and English curated by my lab.[17] These include the best-known novels from the period as well as well-known nonfiction, including philosophy, essays, and histories. These texts have been sufficiently cleaned so that they are subject to minimal transcription errors and also broken out by point of view so that we can control, for example, for third-person novels when comparing to historical narratives.

The second collection is a much larger sample of nineteenth-century writing in English consisting of 21,158 documents, both fictional and nonfictional, that are drawn from Ted Underwood's research using the Hathi Trust archives.[18] This group, whose contents are much less well understood, allows us to test our results between a canonical subset and a much broader group of writing from the same period. The third component consists of a collection of about 6,500 novels drawn from both the Stanford Literary Lab's nineteenth-century novel collection and the Chicago Text Lab's collection of twentieth-century novels. Together, these collections allow us to examine diachronic shifts in the novel's vocabulary. Finally, the fourth component consists of a collection of 800 contemporary novels and nonfiction published within the past decade that has been curated by my lab.[19] This gives us some traction on the extent to which the effects we are seeing in the past continue to hold in the present.

If the sample of texts tested is one key foundation of the insights that can be gained, the features used provide the second foundation. The way we model documents conditions what we can know about them. Different features will allow for different insights. The features that I will be exploring here will be drawn largely from the Linguistic Inquiry and Word Count Software (LIWC) developed by the linguist James W. Pennebaker.[20] LIWC consists of eighty different features that range from the identification of syntactic and grammatical features like the use of punctuation, prepositions, verb tense, and pronouns to higher-level cognitive phenomena like social, perceptual, or emotional processes that are based on fixed lists of words.[21] The dictionaries have been tested and validated on human subjects, the results of which are available for review.[22] As with all lexicon-based approaches, there are open questions as to the semantic coherence of a given category. Are all the words in the "insight" dictionary really indicative of moments of cognitive insight in novels? Or do all instances of *I* mean the same thing?[23] The interpretation of these results thus needs to be handled with a good deal of caution. In particular, care needs to be taken in how the categories are understood *as categories*, with emphasis given to assessing the semantic coherence of these categories within the novels. When we drill down into the individual features, what do we find?

At the same time, it is also important to emphasize the benefits of such lexicon-building approaches for the computational study of literature. First, it provides an important means of dimensionality reduction. If we treat every word in a text as a variable, we can easily end up with thousands, if not hundreds of thousands, of dimensions for a collection of texts. This has serious negative consequences for the statistical validity of any comparisons made between the groups (many more variables than observations runs the risk of "over-fitting," meaning there are too many particular, and potentially meaningless, ways to tell the difference between groups of documents for the model to be generalizable). LIWC collapses this space into a much smaller range of more general semantic and syntactic categories. But unlike unsupervised approaches like topic modeling or principal component analysis, to name two of the more popular approaches that also help reduce the number of overall dimensions in a model, the families of words being tested here (call them "topics") are independent of the data being tested. In a topic model, labels are supplied after the fact, and are thus very open to the problem of confirmation bias. And as we saw in the last chapter, they are not necessarily any more semantically coherent than a supervised list of words. With LIWC, or any other dictionary-based approach, we start with prior assumptions about linguistic categories and test their presence within given text collections. The externality of the words from the collections they are meant to study allows us to test beliefs independently of the collections themselves. While neither approach is perfect, in both cases we are moving between individual words and the ideas those words are thought to embody. What is ultimately at stake is the confidence of a *model* to approximate some underlying textual phenomenon. The key is making as explicit as possible how we move between these different levels of analytical scale, that is, how we connect the conceptual, the lexical, and the theoretical.

LIWC can thus give us an initial range of intuitively meaningful interpretive categories to build on as well as the lexica upon which those categories may in part be based. They should not be taken at face value, but looked into, as with all semantic fields. They offer an alternative to the derived linguistic fields that I explored in the previous chapter. No one method suits all questions. Because the dictionaries are transparent in LIWC, users can refine or alter the dictionaries as they see fit, as I have done on occasion here. They can also be combined with other, more customized, features, as I will be doing in the chapters that follow. Building more sophisticated feature sets is ultimately one of the central research challenges of the field.

The Coherence of Fictionality

The question that I want to begin with is: How coherent is fiction as a type of writing? Are there indeed no syntactic or semantic properties, as Searle contends, that allow us to predict whether something is intended fictionally? Is fictionality exclusively a function of communicative context, the intentionality of the writer, and the belief system of the reader? Or are there features that appear with a high degree of regularity in fictional texts that do not appear in nonfiction, such that even a computer can make accurate guesses as to the nature of the text?

In order to answer these questions, I will use the process known as machine learning to see how accurately a computer can predict a text's given class (I will be using the learning algorithm known as a support vector machine [SVM], which is applied in many text classification scenarios).[24] For those not familiar with this process, a learning algorithm is "trained" on features found in a set of documents for which the classes are already known and then asked to predict which class a group of texts belong to that it has not seen. In this case, I train the algorithm using the LIWC features discovered in a given set of documents and use a process of tenfold cross-validation to make predictions on whether a document is a work of fiction or nonfiction. What this means in practice is that I randomly divide the corpus into a 90–10 split ten times, where 90% of the documents are used to train the algorithm and the unseen 10% are used to test its reliability. (The folds function in the kernlab package ensures that the folds are equally divided between the two categories relative to the overall data.) Doing this ten times allows us to gain a full view of all the documents in the collection, as each document has an opportunity to be in the test set.

Table 4.2 Classification results for predicting fictional texts using tenfold cross-validation

Corpus1	Corpus2	Avg. accuracy (F1)	No. docs
Fiction (EN_FIC)	Nonfiction (EN_NON)	0.94	100/100
English Novel (EN_NOV)	Nonfiction (EN_NON)	0.96	100/100
German Novel (DE_NOV)	Nonfiction (DE_NON)	0.95	100/100
English Novel 3P (EN_NOV_3P)	History (EN_HIST)	0.99	95/86
Germ Novel 3P (DE_NOV_3P)	History (DE_HIST)	0.99	88/75
Cont. Novel (CONT_NOV)	Nonfiction (CONT_NON)	0.96	193/200
Cont. Novel 3P (CONT_NOV_3P)	History (CONT_HIST)	0.99	210/200
19C Fiction (HATHI) (Trained)	Cont. Novel (CONT) (Tested)	0.91	21,158/400

Table 4.2 presents the results of this experiment, showing which two data sets were compared and the average accuracy of the predictions on the unseen data. As we can see, not only are the differences between fiction and nonfiction robust across time and languages, but we can use models built in one time period to strongly predict those of another. In the last line, I train a model on nineteenth-century fiction and nonfiction from the Hathi data set and then test it on the contemporary novels and nonfiction. While there is a clear drop in performance when we use nineteenth-century models to predict twenty-first-century novels, we can still see a relatively high degree of performance at work here (around 91% accuracy). There appears to be a notable degree of diachronic stability to fictional discourse over the past two centuries. Indeed, as we will see below, when we examine features that are more indicative of novelistic writing in particular as a subset of fictional discourse, we generally see these features increase over time. The trans-temporal stability of the novel is complemented by an increase in certain types of novel-specific vocabulary that can be traced back to the nineteenth century.

The Phenomenology of the Novel

If fiction is so predictable, what are the features that make it so? Here I use a rank-sum test to examine which features are statistically distinctive of one group compared to another by observing the overall distributions of each feature in the two groups.[25] The more those distributions differ from one another, the stronger the statistical significance. I then rank the features by the ratios of the median values for each group.[26] The value of doing so is that it preserves information about the overall distribution of a feature in a given population (rather than being driven by a few texts that might have a particular feature in a much higher amount). The disadvantage is that it does not correct for low-occurring features—small features that change considerably will look more important than highly common features that only change slightly. This may not be a disadvantage depending on what one's assumptions are, but it is important to understand the way this assumption is built into the rankings.[27]

To begin, let me review the overall structure of the tables used here to better understand what they can tell us (see table 4.3 as an example).[28] The leftmost column ("Feature") lists the features as defined by LIWC. Some are extremely straightforward ("exclam" refers to the percentage of exclamation marks), while others are more nuanced. "Family," for example, refers to a dictionary of words all related to family members, while "social" relates to words having to do with social experience, which can include pronouns (a choice that effectively duplicates the pronoun categories because they are so much more common than other words). The former is arguably much more straightforward than the latter, and thus we need to be cautious when we encounter a dictionary that is more semantically ambiguous (though even a single word like "you" may have different kinds of functions in novels). The second column ("Category") lists the category to which the feature belongs, a slightly more general framework for understanding the individual features. The next two columns (Fiction %, Nonfiction %) present the median frequency of that feature in each corpus as a percentage of all words. This allows us to see which features are more prevalent relative to other features.

Because percentages are somewhat opaque in terms of a reader's experience, I will generally be translating these numbers into page and work equivalents in the discussion that follows. This allows us to imagine our way into a reader's experience and surmise which features occupy more of a reader's attention. Exclamation marks, for example, comprise on average about 0.45% of a given work of fiction in the nineteenth century. If we assume an average novel length of about 100,000 words (or 500 words per page across 200 pages), this means that there is one exclamation mark for about every 200 words, or 2–3 per page, or roughly 500 total per novel. Personal pronouns, on the other hand, occur about 10% of the time in fiction, which means once every 10 words, or *50 times* per page (and 10,000 times per novel).

Table 4.3 The top ten features with the greatest increase in fiction compared to nonfiction. Values represent median percentages in the nineteenth-century canon collection.

		Fiction vs. Nonfiction 19C canon (English)				
Feature	Category	Fiction (%)	Nonfiction (%)	Ratio	Sample rank	Hathi rank
exclam	linguistic	0.40	0.04	10.00	1	2
you	linguistic	1.34	0.16	8.34	2	1
q-mark	linguistic	0.41	0.08	5.13	3	6
I	linguistic	2.41	0.49	4.92	4	7
quote	linguistic	2.59	0.65	3.98	5	5
assent	social	0.12	0.03	3.83	6	4
family	social	0.56	0.16	3.58	7	10
hear	perception	1.15	0.38	3.01	8	9
shehe	linguistic	4.86	1.90	2.56	9	8
ppron	linguistic	10.73	5.01	2.14	10	14

p-value < 0.0001

The fifth column, "Ratio," lets us see how much more prevalent the feature is in one collection over another by comparing the ratio of its median value in the two collections. Exclamation marks appear almost ten times as often in fiction as in nonfiction. This is a massive difference, but we are still only talking about something that occurs relatively infrequently compared to other features. Personal pronouns, on the other hand, only appear a little more than twice as often in fiction (still a very large difference), but this increase is based on a much larger linguistic aspect of texts. Twice as many pronouns means roughly 5,000 more pronouns per work, or about 25 more *per page*. While I privilege ratio here in my interpretation of the results, we will want to keep our eye on both of these aspects, from the overall prevalence of the feature to the relative increase from one population to another.

Finally, the columns "Sample rank" and "Hathi rank" refer to the respective rankings of a given feature in the small canonical data set of about 100 novels in English versus the much larger collection of over 9,000 documents in the Hathi Trust collection. The idea here is to try to understand the extent to which the smaller sample serves as a decent approximation of the much larger set.

As I proceed, I will be translating these tables into word clouds in order to render the information in a more digestible way. Word size refers to the ratio value, where larger equals a higher prevalence in fiction, while color (words in black) refers to a particular category of interest.

CHAPTER FOUR

4.2   Distinctive features for nineteenth-century fiction. Words in black correspond to linguistic categories such as punctuation, pronouns, and verb tense.

Several of the tables are included in the appendix, and all are available in the supplementary data for further review.

Beginning with the baseline comparison of fiction and nonfiction writing using both our canonical sample and the larger collection of Hathi Trust writings from the nineteenth century (fig. 4.2), we see how the features that are most indicative of fictionality are driven by dialogue—exclamation marks, question marks, quotation marks, first- and second-person pronouns like *I* and *you*, assent words like *yes*, *okay*, and *oh*, and finally the word *said* (which is labeled as an auditory verb by LIWC). Importantly, we also see very strong alignment between the nineteenth-century sample and the larger population of Hathi Trust documents, with some notable exceptions around the "social" category and potentially *family*, *home*, and *ingestion*. If we compare these groups directly, we see that only *family* and *ingestion* are somewhat inflated in the canonical sample (by about 10–15%).[29] In other words, while there are interesting variations that are worth exploring, on the whole, the smaller sample does a good job of capturing the same information as the larger collection.

Taken together, these features suggest a relatively unambiguous way in which fictional writing has a uniquely dialogical construction when compared with nonfiction. While this may not be "news," it does help us build a taxonomy of the distinctions that make this kind of writing socially significant. Imagining people talking to each other appears to be one of fiction's primary cultural functions.

Indeed, imagining people *as people* may be fiction's most important role. If we remove dialogue from the sets above, including the pronominal expressions that accompany them (she said, he cried, etc.), we see how third-person pronouns emerge as one of the strongest indicators of fictionality, along with references to family members and bodies (fig. 4.3, table A.6).[30] There is over a threefold increase in the average number of she/he pronouns in fiction versus nonfiction outside of dialogue, with just these two words alone accounting for more than 5% of all words in the text (or roughly 5,000 instances for a medium-length novel).

This is especially remarkable considering that on average, works of history, for example, use considerably more proper names than works of fiction (an estimated more than twice as many).[31] The lower number of people in fiction is compensated for by a more expanded durational existence on the page for which pronouns become key linguistic markers. People seem to have more extended identities in fiction, though this is not necessarily to be confused with a more "expansive" identity, that is, one that is more semantically rich. The pronominal frequency of characters is not the same as the linguistic diversity surrounding these characters (a point to which I will turn in my next chapter). Nevertheless,

4.3 Distinctive features for nineteenth-century fiction with dialogue removed. Words in black correspond to social and biological categories.

this gives us a first indication of the ways in which fiction performs the process of identification as a repetitive and extensive act of naming the same person.

The prevalence of family and friend vocabulary in fiction also suggests what type of people are more distinctive of the genre, just as the setting of home gives us an idea of where they are most active. Broadly speaking, when we read fiction in the nineteenth century, what is novel, that is, different from other kinds of texts that purport to be about real things, is a focus on family and the familiar. Travel, adventure, work—these can be experienced elsewhere in ways that documentation of family life and the extended agency of individuals cannot. Family is the dominant social imaginary of nineteenth-century imaginative writing.

The stakes of this attention will become even clearer when we focus on a particular type of fiction (novels with external narrators) and a particular type of nonfiction (history writing) (fig. 4.4, table A.7). What rises to the top here (seen in the top cloud) are a host of perceptual categories (seeing, hearing, feeling) that construct the phenomenological reality of an experiencing individual (all more than three times more likely in fiction in the nineteenth century).[32] And the greater prevalence of body words (again about 2.5 times higher in the nineteenth century) gives us an indication of where that attention most often lies. It is knowledge, not just of otherness, but of another *embodied* individual, that most consistently frames the epistemological horizon of the novel from a quantitative point of view.

Interestingly, when we look at the German and contemporary data sets, we see some slight nuances to this story (fig. 4.4, middle and bottom).[33] Without being able to reliably remove dialogue from the German texts, those dialogical markers of pronouns and punctuation marks dominate, but just beneath these are once again the body and sense perception words at work (with some more emphasis on affect than in the English corpus). In the contemporary novels, we see how embodiment has become as important as sense perception—it is now over 4.5 times more likely to be present in fiction than in history. In the contemporary novel, there is an even stronger focalization effect taking place between sensory experience and the observed human body.

These results pose an interesting challenge for "theory of mind" approaches that argue that fiction's primary purpose is the enactment of another human consciousness.[34] While we will see an area where this hypothesis does make sense in the next test, in terms of understanding the novel's distinctiveness compared to nonfictional writing, the mind-body distinction that underlies theory of mind models does not hold up

body
family shehe
anx see social
bio   ppron
sad hear ingest
friend feel home
percept discrep
sexual

humans
othref hear
senses family
sleep qmark
posfeel i assent
you   I sexual
we self friends
see body
physical
pronoun

assent
ppron shehe
sexual qmark period
feel you family
home   see
social i body bio
present   hear
ingest percept
pronoun

4.4 Distinctive features for third-person novels compared to histories for nineteenth-century English (top), nineteenth-century German (middle), and contemporary English (bottom). Words in black correspond to the categories of sense perception and embodiment.

CHAPTER FOUR

percept **social**
excl **present** **future**
verb **assent**
hear
feel **qmark** past
insight **period** **tentat**
see
discrep shehe
negate adverb
auxverb

4.5 Distinctive features for nineteenth-century novels compared to other fiction from the period. Words in black correspond to words related to cognitive processes.

well in light of the novel's strong emphasis on sensorial input and embodied entities. The sensual experience of a sensing being: this is what appears to be uniquely reiterated in the imaginative work of novelistic writing when compared to the nonfiction of historical writing.

As a final way to understand, and bring into sharper relief, the significance of what I am calling the phenomenological orientation of the novel, I compare the nineteenth-century novel with a particular subset of fiction that excludes novels published during the same time period (fig. 4.5, table A.8). Non-novelistic fiction in this case refers to a broad mixture of fictional writing that would have been very present to nineteenth-century readers, including classical epics translated into prose (*The Iliad, Odyssey, Edda, Nibelungenlied*), classic works of prose fiction (*The Tale of Genji, The Decameron*, King Arthur tales, and Rabelais), fairy tale collections from around the world (drawn from Irish, German, Danish, Japanese, and Indian sources), contemporary novella collections (novellas by Hoffmann, Tolstoy, Dickens, Maupassant, Hawthorne, and Washington Irving), as well as a variety of "tales" collections (*Tales of Former Times, Tales of Domestic Life, Moral Tales*). This data set is meant to represent a range of prose fiction that would have been widely read and known to nineteenth-century Anglophone readers but would not have been considered a "novel." While the material dates from different epochs, the publications (and translations) are all contemporaneous with the period as a whole.

Three interesting features initially stand out. First, the ratios are much lower when compared with nonfiction (ranging from 65% to as low as 10% increases, as can be seen in table A.8, which is far below the 200–400% increases we were seeing above). While these groups are similarly well differentiated when compared to nonfiction, when compared to each other the overall distinctiveness drops considerably. If we run the same classifier as above, we can predict novels with about 68% accuracy, which is just below the threshold of statistical significance ($p = 0.018$). If we use a slightly larger collection of novels from the Hathi Trust collection (428 to mirror "other fiction"), accuracy will increase slightly, to 74% ($p = 7.23\text{e-}05$). While this is well above random, it is nevertheless considerably lower, for example, than the ability to predict novels from different genres. As Ted Underwood has shown, it is possible to predict detective fiction and science fiction across a 150-year span with between 88 and 90% accuracy.[35] And as I've shown elsewhere, using a similar-sized collection of contemporary novels, we can predict romances versus more general popular novels with about 98% accuracy, science fiction with about 87% accuracy, and mysteries with about 85% accuracy.[36] The broad category of "other fiction," then, is not as differentiated from novels as particular novel genres are from each other.

Second, while we see some of our more familiar linguistic fictional markers, such as pronouns and dialogue, we also see a new feature in the category of verbs. There are more verbs overall, as well as more varied tenses (past, future, and present, in addition to auxiliary verbs). In other words, there appears to be greater temporal complexity to novels than can be found in fiction more generally. While this deserves its own study, it suggests an initial insight into one of the key ways that novels differentiate themselves from other kinds of imaginary writing in the nineteenth century.[37]

Finally, we also see a new category emerge here that we have not seen before, one that falls under the heading of "cognitive process." These are the dictionaries that LIWC labels "discrepancy," "negation," "tentativeness," and "insight." If we examine the words in those dictionaries that are most distinctive of novels (and here I rank by log-likelihood ratio), we can see the extent to which these are words that tend to mark out moments of self-reflection, doubt, and hesitation—a kind of testing relationship to the world (fig. 4.6, table A.9).[38] It suggests that where fiction

CHAPTER FOUR

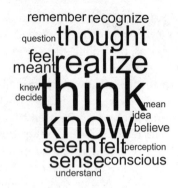

4.6 Distinctive words in novels compared to other kinds of fiction across four categories (moving clockwise from the top left): "discrepancy," "insight," "negation," and "tentativeness."

overall invests in sensorial embodiment, the novel's signature is more oriented toward cognition.

Modal verbs in particular are extremely prevalent here (*could*, *would*, *must*, *might*, and *should*, as well as their negative contractions), and so too is the act of negation more generally (*don't*, *can't*, *didn't*, *not*, *never*, *nobody*). As the presence of "if" suggests, these groups offer different ways of expressing conditionality or even impossibility. At the same time, indefinite words such as *something*, *somebody*, *anything*, and *anybody* are more prevalent, along with a more specific vocabulary of hesitation (*perhaps*, *chance*, *hope*, *possibly*, *guess*, *maybe*, *doubt*, *uncertain*). In between the conditional and the impossible language of the novel, there lies a considerable amount of potentiality—chance, but also skepticism.[39]

Finally, we see how novels are marked by a much stronger use of mental states, captured in major verbs such as *know, feel, think, remember,* and *believe*, along with a second layer of less frequent, but similarly distinctive, complex cognitive verbs such as *admit, ponder, imagine,* and *forgive* (the latter not shown). This is the ground of the novel's reflectiveness, that which binds together doubt and conditionality into a consistent mental state. Indeed, the combination of *seem* and *feel*, both of which appear 30% more often in the novel, gives us a particular indication of what I am calling the novel's phenomenological orientation. Not the world itself, but a person's encounter with and reflection upon that world—the world's *feltness*—is what marks out the unique terrain of novelistic discourse when compared with other forms of classical fiction. It is this combination of sense perception plus cognitive skepticism that seems to bring out the novel's contribution to fictional discourse. The novel professes its uniqueness in the way it offers extended reading experiences of the human assessment of the world's givenness.[40]

The Great Reversal

I have over the course of this chapter tried to support three distinct arguments about the nature of fictional writing since the nineteenth century, with a particular emphasis on the novel. I call these arguments the coherence hypothesis, the immutability hypothesis, and the phenomenological hypothesis, respectively. As mentioned at the outset, I use the term *hypothesis* because these positions are still tentative. They need to be tested with different historical samples, on different kinds of subgenres, using different kinds of features and especially across more cultural spaces. As I have said repeatedly throughout this book, this is just the beginning.

But when we begin to look at the nature of fictional discourse from a quantitative perspective, there does appear to emerge a relatively clear story about its larger social function and the ways it distinguishes itself from purportedly "true" writing. Using the approach of machine learning, we are able to see how coherent fictional writing is when compared to a number of different kinds of nonfictional writing in different periods and in different languages. Seen in this way, and following other work in the field on the coherence of specific fictional genres,[41] notions of the indefiniteness or openness of literature look profoundly overstated. There is an underlying consistency or integrity to fictional discourse that distinguishes its intended (un)truth content from its nonfictional counterparts. While in theory anything can be a novel, it turns out that in practice, when writers signal the fictionality or even novelism of a work, they do so in highly predictable ways.

The *way* they do so, as I have tried to show through the descriptive sections of the chapter, is with an emphasis on what I would call phenomenological encounter. The linguistic investment of fictional writing falls overwhelmingly on sense perception and a sense of human embodiment. For the novel in particular, this will take the form of an increased investment in conditionality and the hypothetical, in a questioning relationship to the world. When seen from a quantitative perspective of where the vocabulary mass of novels or fiction lie, they appear to focus strongly not on things "out there," on a kind of social horizon, but rather with individual encounters and embodied assessments of what is.

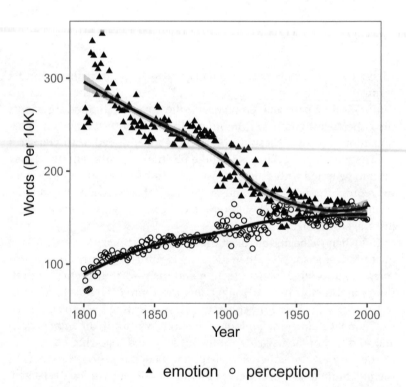

4.7 Frequency of words related to emotions and perception in English-language novels, 1800–2000.

This orientation of fictionality has remained remarkably constant for over two centuries. As we saw in the machine learning section, models trained on nineteenth-century fiction can strongly predict twenty-first-century texts. And yet, where there have been changes, these changes only appear to underscore the phenomenological orientation of fictional writing that I have been arguing for here. As we can see in figure 4.7, the great reversal of novelistic writing over the past two centuries has been the wholesale swap of abstract emotional states for embodied sense-perceptual ones.[42] Whereas the nineteenth-century novel began highly invested in emotionality, by the mid-twentieth century this would drop to levels commensurate with a mid-nineteenth-century investment in sense perception.[43] Similarly, these explicit acts of sensing and feeling would continue to rise, so that by the Second World War they are appearing close to five times on every page we read.[44]

If we compare levels of embodiment between the nineteenth century and the present, this time understood not as sensorial actions, but as pieces of the sensorium, we also see a similar shift in emphasis. Contemporary novels use 65% more terms of embodiment than nineteenth-century novels, 42% more perception-related words, and 37% fewer emotional terms.[45] It is this rise in phenomenological investment that accounts for the increasing schism between contemporary fiction and nonfiction where, as I discussed above, we see a 450% increase in the language of embodiment compared to historical writing, by far the most important "category" of words that marks out the contemporary novel when compared to nonfiction.

The great reversal, then, is a kind of asterisk or caveat within the larger immutability of fictional writing over the past two centuries. If we reduce expression down to just these two dimensions, we see how "feeling" is gradually redefined over time by the novel away from an emotional state to one of perceptual encounter. This sense of the contemporary novel's investment in sensorial embodiment awaits fuller treatment.

FIVE

Characterization (Constraint)

"I believe that all novels begin with an old lady in the corner opposite."
VIRGINIA WOOLF

"The worst thing, it seemed to her, was to be dealing with one version of a person when quite another version existed out of sight." Rachel Cusk's *Outline*, from which this quotation is drawn, is a novel, at once beautiful and eerie, that is concerned with the construction of character. It asks how we make characters from language and in doing so always leave something out. "It is surely not true," says one of Cusk's characters during a creative writing seminar in Athens, "that there is no story of life; that one's own existence doesn't have a distinct form that has begun and will one day end." For Cusk, there is an essential tension between this desire for form—that a character can have a figure or outline—and the sense of incompleteness or hollowness that surrounds the task of encasement—that we never get anything more than an outline. "In everything he said about himself, she found in her own nature a corresponding negative. This anti-description, for want of a better way of putting it, had made something clear to her by a reverse kind of exposition: while he talked she began to see herself as a shape, an outline." Character, according to Cusk, is potentially nothing more than a distinction.

Cusk's concerns with the hollowness of character have a long literary pedigree. As George Eliot would satirically write in *Middlemarch*, "'He has got no good red blood in his body,' said Sir James. 'No somebody put a drop under a magnifying glass, and it was all semicolons and parentheses,' said Mrs Cadwallader." Or later, in the hands of Thomas Pynchon, who writes his protagonist out of *Gravity's Rainbow*: "He's looking straight at Slothrop (being one of the few who can still see Slothrop as any sort of integral creature any more. Most of the others gave up long ago trying to hold him together, even as a concept—'It's just got too remote' 's what they usually say)." Slothrop, the novel's hero, will eventually be dispersed into the novel's so-called "fingervectors." At least since the eighteenth century, novelists have regularly plumbed the feigned contours of their fictional people.

At the same time that novelists have been casting doubt on characters' existence, scholars have been trying to reconstruct them. What characters are and do has been a regular question on the literary critical circuit, despite periodic protests to the contrary that character has somehow been undertheorized. For the Russian formalists, embodied above all in the work of Vladimir Propp, character was primarily a "type," one that served different narrative functions ("the hero is married and ascends the throne").[1] For post-structuralists that came in the wake of Propp, character was nothing more than a rhetorical "effect," one more example of the referential fallacy of naïve readers.[2] Characters were ambiguous semantic bundles, neither types nor individuals.[3] More recent work has drawn on the field of cognitive science to argue that characters are useful tools through which to model "theories of mind," means for learning about and hypothetically experiencing human cognition.[4] Still indispensable is the work of Deidre Lynch and David Brewer, who explore the historically specific uses of character for readers. The afterlife or extra-textual life of characters, so this work has shown, can tell us a great deal about the matter and mattering of characters at different points in time.[5]

This chapter argues that quantity has an important role to play in understanding the nature of characters and the process of what we might generally term characterization—the writerly act of generating animate entities through language. As figure 5.1 shows, and as Alex Woloch has recently hypothesized, characters have a very distinct distributional signature.[6] There is a vastly unequal relationship between the attention granted to a novel's "main" character(s) and everyone else. This is what Woloch terms "the one and the many."[7] While Woloch rightly wants us to see this distinction as crucial to the meaning of the novel as a genre, traditional modes of reading are not well positioned to understand the problem of "the many" that resides at its core. With an estimated 85 characters per novel in the nineteenth century and a conservative

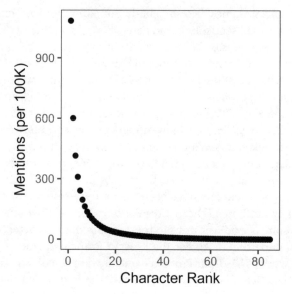

5.1 Frequency of character mentions by character rank in nineteenth-century novels.

estimate of 20,000 novels published during this period in the English language, there are roughly 1.7 *million* unique characters that appear in that one century and one language alone.[8] Even if we condition on main characters, the so-called "one," we are still looking at over 20,000 distinct entities.

At the same time, the process of characterization itself poses a challenge of scale. Not only are there a great number of characters, but there is also a tremendous amount of information surrounding even a single, main character. Like other highly frequent textual features, such as conjunctions or punctuation, characters are abundant across the pages of novels. Personal pronouns alone account for roughly 12% of all tokens, and if one adds in proper names, the number of character occurrences is closer to 16%—or one in every six words! (For reference sake: verbs represent about 14% of tokens, conjunctions about 6%, and periods 8%).[9] Like the abundance of characters themselves, such semiotic abundance poses problems for inherited critical methods. How can we be sure that our claims about "character" are capturing the broad and potentially diverse ways that characters are depicted in novels, this larger mass of fictional beings and what it means to be fictional?

This chapter tries to account for that lost information surrounding character. It uses computation to better understand the language associated with characters, what we might call the "character-text" of novels.[10]

By "character-text," I mean words that stand in a dependent relationship to a character—words used to describe characters—but that do not include dialogue (which would constitute its own text space, the "dialogue-text"). What interests me is the process of *characterization*, the narrative construction of characters prior to their enunciative embodiment. I focus here on what David Bamman, Ted Underwood, and Noah Smith have identified as four core modes of being: what characters do, what they have done to them, how they are described, and what they possess.[11] To what extent do these textual spaces associated with the building and instantiation of characters distinguish themselves from other narrative aspects of novels, and to what extent are they semantically richer or, as we will see, poorer? How "special" are characters in their sense of being (rather than speaking)?

The argument that I will be making, one based on an examination of 7,500 novels and over 650,000 characters from the past two centuries, is that while characters do appear to represent a distinctive semantic space in novels, they are more constrained and homogeneous than other aspects of narration. There appears to be a strong degree of uniformity that exists between characters from different novels, even main characters, just as there seems to be a strong degree of uniformity between characters within the same novel. I have even found that characters tend to look more similar over the course of a novel than other types of nominal things.

Far from serving as a proxy for the individual or individuality, then, the evidence here suggests that the process of characterization is best described as one of stylistic constraint. It aligns the practice of characterization more closely with a character's etymological origins as a sense of being representative, general, or "characteristic." We are always with characters (every six words). But rather than think of them as singularly deep mirrors of ourselves, as unique entities, computation foregrounds the ways we are encountering highly distributed entities, a point that seems to have changed little over the past two centuries. The process of characterization is closer in this view to Roland Barthes's thinking about character as a "transitory site," one that organizes semantic codes around itself, or more recently, Megan Ward's attention to the *seemingness* of characters, the way they foreground in her words their "blatant artificiality."[12] In this sense, characters carry more than they reflect. The phenomenological orientation of novels that we saw at work in the last chapter is not uniquely imbedded in a single entity, or in the production of singularity. It is distributed across the larger practice of characterization itself. Characterization is the act of building a scaffold of mediation—between characters and between characters and the worlds they inhabit.

Seeing characters in this way suggests that the discourse surrounding characters' complexity, their "semic surplus," in the words of one commentator, or the cognitive scientific emphasis on deep cogitating minds, misses a core aspect of characterization.[13] The refrain of dividedness that has been echoed by novelists over the centuries—that is, of a core duality surrounding characters, between their form and their content—appears to be a more fundamental insight about the nature of characterization. The person—the entity that we endow with coherence and agency—lies in many ways beyond the text.[14] As Barthes once suggested, "To read is to struggle to name."[15]

And yet, at the same time that we see this more global coherence of character—the way characters are distinct for not being distinct—we also see a certain amount of local difference and variability between characters at a more specific level. In the second section, I discuss a character-feature tool that we have developed in my lab that tries to identify different practices of characterization. Despite the more global regularity of character, we can still detect local differences between types of writing and types of characters, even if those indices are far weaker and more heterogeneous than the recognition of character more generally. The differing strength of these signals has something important to tell us about the ways readers are adept at endowing otherwise highly uniform entities with distinctiveness and agency, the way small cues carry a large effect. In the second half of the chapter, I use the notion of constraint that surrounds character more generally to explore the rise of interiorly oriented characters in novels. What I have found is that female protagonists by female novelists tend to be far more cogitative and perceptually oriented than their male counterparts. The lexical constraint surrounding characters is translated in the hands of an emerging group of women writers in the nineteenth century into an exploration of social constraint, a new form of agency predicated on thinking and observing.

When we talk about the rise of deep characters, then, about interiority as a hallmark of the modern novel, it is crucial to acknowledge the work of feminist scholars who long ago identified the gendered aspect of this arc.[16] While not belonging exclusively to the domain of women and their heroines, as a number of scholars have pointed out, it is clear that one of the major contributions of women writers of the nineteenth century, beginning most notably with Ann Radcliffe and Jane Austen, is the privileged development of characters whose inner lives of thought and feeling come to stand for novelistic character more generally. The phenomenological orientation of the novel that I discussed in my previous chapter is strongly tied with a particular type of characterization developed by a particular group of writers. Leaving women out of the story of the rise of the novel means we are overlooking one of the major cultural contributions made by women's entry into the literary marketplace.

The value of a computational approach to character is not simply that it allows us to validate earlier insights across a larger sample of writing, though to be sure this too is important. It also allows us to specify more clearly the contributions by women novelists, to identify the different ways that interiority took shape and was put to use. As we will see, that interiority was not solely tied to a notion of introspection or self-reflection, but also contained an aspect of futurity and anticipation. The introverted heroine was very often impressed by a range of intense and often troubling emotions, anxiety being a kind of catchall term for this verbal field. It suggests that the point of this characterization was not some form of retreat from action—it was agential to the core—but rather there to posit a profound sense of potentiality, a cognizance of the risks of something new to come.

Perhaps even more interesting than the identification of the emergence of a particular character profile is the way computation shows us how this gendered orientation of character in the nineteenth century *changes* over the course of the twentieth century to assume very different connotations. While we have talked about the rise of female subjectivity in the novel, we have missed where this particular form of characterization goes after its classical realist phase. As we will see, the phenomenological orientation of character in novels by women is transferred to a new, largely male, and largely techno-scientific framework in the twentieth century. The observing cogitating heroine turns by the twentieth and twenty-first century into the (mostly male) self-reflexive subject imbedded within a technological landscape. The social constraints of sexual politics morph into concerns with the technological conditions of contemporary life. Such are the curious circuits charted out by computational characterology.

The Character-Text

How do we go about building a model of character, of establishing the character-text of a novel? Consider the following passage taken from the closing chapter of *Pride and Prejudice*, in which only words associated with characters are kept:

her feelings Mrs. Bennet rid her deserving daughters she visited Mrs. Bingley talked Mrs. Darcy I I say her family her desire her children make her sensible amiable well-informed woman her life her husband who might felicity she nervous silly

Here is the same passage without character-words:

Happy for all maternal was the day on which got of two most. With what delighted pride afterwards visited, of, may be guessed. wish could, for the sake of, that the accomplishment of earnest in the establishment of so many of produced so happy an effect as to a, , for the rest of; though perhaps it was lucky for, not have relished domestic in so unusual a form, that still was occasionally and invariably.

And here is the passage fully reconstructed:

Happy for all her maternal feelings was the day on which Mrs. Bennet got rid of her two most deserving daughters. With what delighted pride she afterwards visited Mrs. Bingley, and talked of Mrs. Darcy, may be guessed. I wish I could say, for the sake of her family, that the accomplishment of her earnest desire in the establishment of so many of her children produced so happy an effect as to make her a sensible, amiable, well-informed woman for the rest of her life; though perhaps it was lucky for her husband, who might not have relished domestic felicity in so unusual a form, that she still was occasionally nervous and invariably silly.

In order to understand characters, we first need a way of identifying them in the text and then identifying words associated with them.[17] To do so, we begin by automatically labeling characters, a much more challenging task than it might at first seem. First, all of the various aliases of characters need to be identified—that is, words that stand for characters, such as pronouns, proper names, but also nicknames—and then those aliases need to be grouped into distinct sets, with each set referring to a single character. Characters are by no means straightforward linguistic entities. Indeed, the closer you look, the more complicated they start to appear. They can have multiple proper names and nicknames (Slothrop, Tyrone, Mr. Slothrop), just as pronouns can refer in various directions to different characters. While the best-performing systems for resolving pronoun and alias references to characters are currently about 86% accurate, we have found that adult readers have little problem with this task.[18] It sheds light on just how adept our brains are at linking up all of this distributed information in order to mentally construct a distinct entity—or how much effort must be invested to train readers to be able to perform this task.

Table 5.1 Words in a dependent relationship with characters in a sample passage from Virginia Woolf's *To the Lighthouse*

Character	Dependency (captured)	Dependency (missed)
her	this, second	smiling, flashed, idea
William	marry	
Lily	marry	
she	took, mixture, measured, it, leg	stocking
James	measured, leg	

Once character occurrences have been identified, the second step is to use what is known as a dependency parser to identify the words that are in a dependent relationship with characters.[19] The simplest relationships that exist in a dependency tree, for example, are between subjects and verbs or adjectives and objects. Part of the meaning of a character is what he or she does in a sentence (the verb or verb phrase), just as part of the character's identity will depend on any modifiers surrounding her (adjectives). Naturally, there is a syntagmatic bias to such a scheme—all words surrounding a character can potentially bear on the character's meaning, even if they are not in a dependent relationship with the character.[20] But dependency is useful because of the way it provides a more limited field of reference. It gives us a more conservative estimate of vocabulary associated with characters. Table 5.1 provides an example of how the Stanford Dependencies parser performs on a sample sentence from Virginia Woolf's *To the Lighthouse*:[21] "Smiling, for an admirable idea had flashed upon her this very second—William and Lily should marry—she took the heather mixture stocking, with its criss-cross of steel needles at the mouth of it, and measured it against James's leg." As we can see, the Dependencies parser has what we would call high precision—when it identifies a word as being dependent on another word, it is almost always right. However, in complex literary sentences, it will miss some words (i.e., lower recall).[22]

Taking all of these steps together, we can create a table where for every character in a novel we extract the words in a dependent relationship with that character and the number of times those words appear. (If we think of this in terms of the vector space models discussed in the introduction, characters are now equivalent to "documents" and the dependency words allow us to understand the semantic distributions surrounding them.) Table 5.2 gives us an idea of what this looks like in practice across a variety of categories. One can already begin to intuit the high degree of homogeneity between these vectors just by looking at

Table 5.2 Most common words associated with characters in two different genres and three authors

Bestsellers	Prizewinners	Austen	Dickens	Woolf
say	say	say	say	say
have	have	have	have	think
ask	see	see	hand	look
look	tell	be	return	see
tell	look	think	look	hand
see	hand	CHAR	CHAR	have
CHAR	ask	do	cry	CHAR
hand	know	sister	take	go
take	go	feel	go	feel
know	think	go	head	ask

the tables. Characters tend most often to *speak* and to *have*, just as they often *look* or *see* things. Body parts like the *face* or *hands* reliably serve as sites of focalization when describing characters, just as other characters are often associated with characters (denoted by "CHAR" in the table). These are the transformations that will underpin the observations that follow.

Character Constraints

The first question I want to explore is how informationally rich characters are in novels. Before we move to a specific study of characters, it is important to understand how characters function as linguistic entities. How much surprise and diversity do they bear in their construction? We often make assumptions about the richness of characters. But how much of this is related to how they are depicted according to the framework outlined above: in terms of what characters do, what they have done to them, how they are described, and what they possess? Do we see any significant differences across genre or historical time?

Once again we can turn to machine learning to begin to answer these questions, only this time instead of simply comparing different data sets using the same features, we can compare different data sets using *different sets of features as well* (table 5.3). How much better or worse can we predict which genre a novel belongs to when we use the language of characters versus the non-character-text? According to this experiment, the more predictive the language of characters is, the more informationally rich characters can be thought to be.

What we see here are the various levels of accuracy in predicting a novel's genre depending on the feature set used, where genre can variously refer to things like literary genres such as "mysteries" or "science fiction," groupings of cultural capital like "bestsellers" and "prizewinners," or historical periods such as "Victorian" and "Romantic" or "twentieth century" or "nineteenth century." The diversity of classes allows us to test characters across a range of different types of groupings as well as different sizes of data sets. The columns refer to the different feature sets used, where "All" equals all words in novels, "Char" represents only words from the character-text of novels, "NonChar" captures the non-character-words, and "NonCharAdj" is the same as "NonChar," only reduced down to the same number of variables as the character set to make it a more equal comparison.

What is interesting here is the way relying on the character-text does not do anything special. It not only depresses accuracy when compared to using all words, it also does not outperform the non-character-words, and in most cases underperforms them. If we remove characters from novels, in other words, we see no decrease in classificatory accuracy. This point bears repeating: *removing characters entirely from novels does not decrease the computer's ability to predict what class a novel belongs to.* While this may be a commentary on the robustness of classification algorithms, it also suggests just how unessential characters are to detecting the kind of writing one is reading.

In a second experiment, I try to capture how similar characters are to each other *between* novels by examining the distributions of words associated with them. Is it the case that characters tend to look more like each other than the novels from which they are drawn? How important is authorial style in accounting for the uniqueness of characters?

To do so, I use a measure called Kulback-Leibler divergence, which calculates the amount of information lost when one probability distribution

Table 5.3 Results of classification tests using machine learning to compare the predictability of different genres using three different sets of features: all words, only character words, and no character words. Values represent average F1 scores using tenfold cross-validation.

Genres	All	Char	NonCharAdj	NonChar	Documents
Contemporary (6x)	0.73	0.66	0.66	0.73	1,184
Bestseller-Prizewinner	0.83	0.76	0.78	0.8	400
Romantic-Victorian	0.93	0.87	0.89	0.94	288
Romantic-18C	0.96	0.9	0.99	1.00	84
19C-20C	0.99	0.97	0.98	0.99	6,646

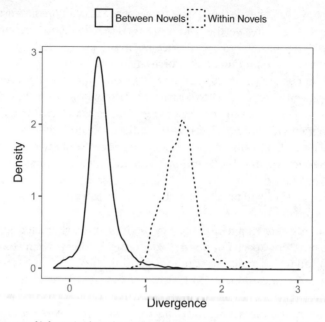

5.2   Amount of information lost when comparing the semantic profiles of characters between novels versus comparing characters to their own novels.

is approximated by another. In our case, the character-texts of novels can be thought of as probability distributions (the likelihood of a given set of words occurring), where we are asking how much information is lost when we approximate this distribution by those of either the non-character-text from the *same* novel or the character-text of *other* novels.[23]

As we can see in figure 5.2, there is significantly more divergence between characters and the novels from which they come ("within novels") when compared to the divergence between characters from different novels ("between novels"), even when we remove words related to dialogue (*said, answered, replied,* etc.). In other words, there is comparatively little information lost when we substitute one character for another from different novels, something that holds true across all classes and time periods.[24]

If characters are more similar to each other across novels than the novels from which they come, are they also more similar to each other in the *same* novel when compared to other kinds of parts of speech, such as nouns? In other words, do characters appear more semantically homogeneous to each other when compared to other nominal things?

In order to test this, I take the average similarity between character vectors for the top five characters and compare this distribution to the average semantic similarity between the top five most frequent nouns per novel.[25] How much semantic range is there on average between these different types of entities in a given novel? Once again we see how characters appear to be significantly more semantically constrained than the most common nouns in a given novel, even when we remove verbs of communication (table 5.4). It is important to note the considerably higher standard deviations in the comparisons of characters to each other (versus nouns), suggesting that there is a much broader range in how homogeneous characters appear relative to each other when compared to the semantic profile of leading nouns in the same novels. The latter tend to vary little across novels in a given genre, while character similarity shows considerable variation. Said another way, the semantic differences between characters vary considerably more across novels than those of nouns.

Interestingly, we also see some patterns emerge at the level of genre. Bestsellers have the most homogeneous characters, at least for the top five, while nonfictional writing has the least. The further forward we move in historical time, the more homogeneous characters also appear to become. The hierarchy of genres recapitulates historical time, with one glaring exception, that of Romance, which has the least homogeneous main characters of all contemporary genres.

Table 5.4 Average similarity between the top five characters of novels (*Characters*) and the top five most common nouns from the same novels (*Nouns*)

Genre	Characters	Nouns	
	Mean ± SD	Mean ± SD	p-value
Bestsellers	0.55 ± 0.15	0.10 ± 0.04	1.11E-105
Mystery	0.52 ± 0.13	0.09 ± 0.04	2.24E-114
SciFi	0.52 ± 0.18	0.08 ± 0.04	4.53E-89
Young-Adult	0.50 ± 0.15	0.07 ± 0.03	2.15E-88
Prizewinners	0.50 ± 0.18	0.10 ± 0.04	8.72E-87
20C	0.47 ± 0.18	0.09 ± 0.04	0
Victorian	0.46 ± 0.20	0.12 ± 0.06	1.10E-84
19C	0.45 ± 0.19	0.12 ± 0.05	0
Romance	0.38 ± 0.15	0.07 ± 0.03	1.42E-76
Romantic	0.35 ± 0.19	0.11 ± 0.06	7.14E-32
18C	0.33 ± 0.19	0.10 ± 0.07	3.78E-10
Nonfiction (21C)	0.22 ± 0.15	0.10 ± 0.04	3.60E-25
Nonfiction (19C)	0.14 ± 0.09	0.11 ± 0.07	0.0065

The final test I want to offer here considers the relationship between characters over *narrative* time, once again comparing them with nouns. If we look at the words associated with characters at different points in a novel, do we see greater degrees of semantic variation than we would with nouns more generally? By comparing the distributions of words surrounding characters and nouns in the first half of a novel with those of the second half, we can gain a rough estimate of how much semantic change surrounds a character (or noun) across two different time spans in the novel. For there to be significant difference to the character's identity, the assumption is that the words associated with that character—what she does, possesses, has done to her, or how she is described—will also have to change in noticeable ways. The character's context should tell us something about new kinds of identities surrounding that character. The degree of change should outperform some kind of control, in this case the amount of semantic variation surrounding nouns more generally.

And yet once again we find that characters tend to experience *lower* amounts of semantic variation over the course of a novel, even when we remove verbs of communication. We found that characters exhibit significantly higher amounts of semantic similarity between the first and second halves of a novel than nouns (character median = 0.65, noun median = 0.53, $p < 2.2e-16$).[26]

When taken together, these tests give us a portrait of what I would call the constraint of character. When we consider a greater number of characters and a greater amount of information surrounding characters, we see how characters appear to impart significantly less information than we have traditionally assumed. They contribute little to no classificatory significance in predicting the identity of novels; they are much more similar to each other across novels than to the language of their own novel; they are significantly more semantically homogeneous as a group than nouns are to each other; and finally, they even tend to change less over time than nouns, pushing against the notion of "development" as one of the core ways characters assume meaning in novels. Character in this sense is largely a stable narrative infrastructure, one that must be imaginatively filled in by the reader. According to the results here, "character" is best understood in Cusk's evocative terms as an outline.

Thinking-Feeling Heroines

And yet when I read characters, I also feel difference. I know intuitively that there are differences to characters, and those differences matter to me as a reader. That "anti-description" that for Cusk describes one character's sense of relation to another often comes through when I read, sometimes more, sometimes less. Within the larger sea of homogeneity, there are smaller differences that seem to matter, that make Mr. Ramsay distinct from Mrs. Ramsay, Josef K. from Billy Budd, or Emma from Ruth. If overall characters are unique in their high degree of homogeneity—unique for not being unique—the question is how to understand the meaningful differences that exist at a more local level that accord with readers' sense of difference.

Rather than look at single characters, however, my interest here is with a more mid-level understanding: are there meaningful *groups* of characters out there, and if so, what does this grouping have to tell us about the history of characterization, and by extension the history of the novel? Within the greater condition of linguistic constraint, are there features that certain characters share, and if so, what do these profiles (again we come back to Cusk's notion of character as outlines) have to tell us? In returning to the question of *type* with which the formal study of character began, my interest lies in understanding the cultural work that such groups might perform, whether we understand them as types, profiles, outlines, or figures.[27]

To this end, my collaborator Hardik Vala and I have developed a character-feature tool that tries to assess the practice of characterization according to 26 different dimensions. The aim is to better understand the nature of different kinds of characters and be able to account for the qualities that bring these groups together. Our features try to address a variety of aspects of character, from questions of agency, distinctiveness, centrality, and behavioral modality, down to descriptive qualities such as embodiment, affect, abstraction, and object-orientedness.[28] The tool's goal is to reimagine "type" not according to Propp's "functions," but as a shifting suite of behavioral, descriptive, and agential *positions*.

One fundamental way of grouping characters that I will test here is to place them along axes of outward versus inward orientation, whether a character is more defined by other-directedness, that is, interactivity and sociability, or self-directedness, that is, thinking and feeling. Such a framework falls along one of the more popular ways of understanding human types, between extroverts and introverts (and now ambiverts). The point here is not to reproduce social scientific models in fictional form, but to ask what cultural work identification with such distinctions produces. This question is particularly salient because it applies the theory about the constraint of character to an analysis of character itself: it asks whether constraint—understood here as a lack of sociability—becomes

CHAPTER FIVE

High Cogitation Low Sociability	High Sociability High Communication (Extrovert)
High Cogitation High Perception (Introvert)	High Perception Low Communication

5.3 Four-quadrant model of character type by introversion and extroversion.

a meaningful marker of character at a particular point in time, whether the very linguistic conditions of characterization become the grounds of imagining a new kind of character.

In order to do so, we can plot characters from 419 classic nineteenth-century novels drawn from the NOVEL750 collection along two sliding scales that compare notions of sociability and communicativeness with the actions of cogitation and perception (fig. 5.3). The more sociable and communicative a character is relative to his or her level of cogitation and perception, I argue, indicates the degree of extroversion on display for a given character. Sociability is measured as the number of instances in which a character appears in the same sentence as another character, while communicativeness is measured as the percentage of times that a character engages in some form of dialogue (captured through a dictionary of 267 verbs of communication). Cogitation refers to acts of thinking and mental reflection, while perception is informed by actions of sense perception. Both are measured using two separate dictionaries, the former consisting of 196 verbs drawn from the "insight" dictionary in LIWC (which contains words like *accept, believe, think, know,* and *feel*) and the latter consisting of 79 words related to sensory experience that were tested in the previous chapter (*look, see, touch, hear, bite, smell, taste, sound, observe,* etc.). To arrive at the ratios between the extroverted and introverted behaviors, the values of each category are subtracted along their respective axes (sociability-cogitation, communication-perception) and scaled accordingly.[29]

What we find when we do so is a relatively strong gender identification along these axes of introversion and extroversion (fig. 5.4).[30] If we plot main characters from classic Romantic and Victorian novels of the nineteenth century, the characters that fall most distinctively in the introvert quadrant (high cogitation/perception, low sociability/communicativeness) are not only female characters, but are written by female writers. At this quadrant's outer bounds lies Jane Austen, demimonde of the female canon. There is a strongly gendered orientation, in other words, toward writing protagonists who are distinctive in their prevalence of thinking and feeling, one that is captured most strongly by Austen's heroines. Perhaps not surprisingly, we find that Emma is one of this type's most pronounced embodiments.

Some of this effect may of course be due to the relationships between extroverted and introverted behaviors, which are not necessarily

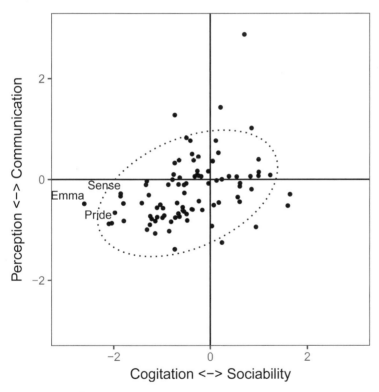

5.4 Ratio of introversion to extroversion in main characters in nineteenth-century novels. Only novels with female authors and female protagonists are shown.

dependent on one another. To avoid this confusion, we can also just focus on the degree of cogitation and perception surrounding main characters in nineteenth-century novels. When we do so, we see how women novelists do indeed tend to write significantly more introverted protagonists than men. The mean level of cogitative-perceptual vocabulary for women is 13.4%, while for men it is 11.8%, meaning a reader will encounter 7–8 more words per page related to thinking and feeling under these conditions.[31] This difference, however, is largely due to higher amounts associated with female protagonists. Male protagonists written by women are not significantly more interiorly oriented than those written by men, just as female main characters written by men are not significantly more introverted than male characters written by women. In other words, when men write women, they tend to look like women writing men when it comes to interiority. The only unique group among the four possibilities of gender combinations between authors and characters in terms of introversion are female characters written by female novelists.[32] This is the distinct contribution of characterization that women novelists are making in the nineteenth century.

That there is a femininity associated with a new kind of subjectivity in the novel around 1800 will be familiar to most students of the novel. As Nancy Armstrong writes in *Desire and Domestic Fiction* about this period, "writing for and about the female introduced a whole new vocabulary for social relations, terms that attached precise moral value to certain qualities of mind."[33] Or, as Gillian Brown argues in *Domestic Individualism*, a new sense of individuality is brought to the fore by a focus on the domestic spaces of women, one that is inflected in her words by values of "interiority, privacy, and psychology."[34] And, as Deidre Shauna Lynch has suggested, "it was women novelists who were in the forefront in writing to the period's new techniques of reading," techniques which, as Lynch shows, were largely guided by the detection of greater psychological depth.[35] By the end of the nineteenth century, critics would canonize this model of character as the "psychology of feminism," identifying certain novels as "modern women's books of the introspective type."[36]

For most critics of the novel, whether contemporary or Victorian, this phenomenon has most often been seen in a negative light.[37] For Armstrong, the high degree of subjectivizing of female characters was a sign of the period's investment in aligning gender and sexual difference, making women masters of the home and thereby emphatically underscoring the two-spheres model. Women were agents in so far as they were objects of desire. Subjectivity was just another form of subjugation.[38] As Gillian Brown writes, this process contributed to "identifying self-hood with the feminine but denying it to women."[39] For Lynch, too, the female vanguard of deep characters was part of a larger social reorientation of reading practices, where readers learned to individuate their readings through the detection of highly individual characters. The female subject was caught up in a greater social experiment in producing class distinction, in the manufactured "status awareness" that accompanied deep reading that is still with us today.[40]

These are powerful and trenchant critiques. They show us the double-edged nature of women's writing practices of the time. But in the process of providing such broad-scale critiques, we overlook at least some of the particular nature of what these writers were doing and why that novelty might have mattered. The emphasis on interiority or depth or even "dispossession," in Catherine Gallagher's terms,[41] bypasses the particular profiles that were being constructed around these modes of behavior. At the same time, it also overlooks the ways in which many female novelists did *not* adhere to these norms. Different positions were open to women writers of the period, and they used these positions for differing ends. Computation can ironically allow us to be more particular: it lets us subset by very distinct linguistic behavior (here, actions associated with thinking and perceiving by main characters)—just as it can allow us to subset by distinctiveness itself—those novels that exhibit this behavior above, below, and within the norms of the period. A distributional approach allows us to observe where the bulk of women's writing is located and at the same time see where the differences within that distribution exist.

Table 5.5 shows us those novels that are exceedingly high as well as low in cogitative and perceptual behavior, and table A.9 in the appendix provides lists of novels broken out exclusively by perceptual orientation. Together, they allow us to see which novels are over- or under-invested in which type of behavior. As I mentioned above, we see how Jane Austen dominates this category, but this is also largely due to her extreme investment in cognitive behavior. When we focus on sense perception, we see a very different kind of novel emerging, novels like Elizabeth Stoddard's *The Morgesons* (1862), Louisa May Alcott's *Moods* (1864), and Susanna Rowson's *Charlotte Temple* (1791). And if we go further down the introversion lists to include a broader swath of writers beyond Radcliffe and Austen, we see a much more diverse group.[42] Elizabeth Gaskell's *Ruth* (1853), Charlotte Brontë's *Shirley* (1849), Mary Jane Holmes's *Marian Gray* (1863), and George Eliot's *Scenes of Clerical Life* (1858) all appear in the top fifteen most interiorly oriented novels (and all more

CHAPTER FIVE

than 1 standard deviation above the mean for nineteenth-century novels more generally).

We see how this investment in women's cogitative-perceptual behavior transpires across a broad swath of time, space, and genre. That these behaviors are indeed highly gendered by the turn of the nineteenth century can be seen in the fact that Mary Robinson's *Walsingham*, about a girl who is passed off as a boy, scores uniquely low for a female novelist. In the novel of cross-dressing, gender is performed in *Walsingham* through the relative absence of introspection that otherwise comes to characterize the work of female novelists writing about women.

If table 5.5 gives us some idea of which works are practicing these types of behaviors above or below their contemporary norms, we need to know more about the values associated with them. What kinds of semantic fields are associated with female interiority? To answer this question, we can use the method of collocate analysis to capture words associated with some of the more important keywords related to thinking and feeling that are distinctive of women's writing. What are the semantic contexts of interiority that are distinctive of nineteenth-century fictional women?

To find out, I focus on the four most common keywords (and their cognates) associated with the actions of cogitation and perception: *looking, seeing, feeling,* and *thinking*. The network represented in figure 5.5 illustrates the words that are significantly more likely to appear near these four words in novels written by women with female protagonists.[43] While it can only give us an initial idea of the semantic profiles of these introverted behaviors, the negative connotations of the cognitive dimensions are indeed striking. *Blame, sadness,* and *bitterness* are associated with the act of thinking, while *disappointment, resentment, reproach,* and *anxiety* strongly associate with feeling (though *wonder* and *tenderness* and *gratitude* are also

Table 5.5 Nineteenth-century novels by female authors with the most and least introverted main characters

Novel	Date	z
Austen, *Emma*	1816	2.59
Austen, *Pride and Prejudice*	1813	2.56
Austen, *Persuasion*	1818	2.53
Radcliffe, *Mysteries of Udolpho*	1794	2.01
Austen, *Northanger Abbey*	1818	1.89
Holmes, *Elsie Venner*	1861	−1.19
Rowson, *Mentoria*	1794	−1.24
Robinson, *Walsingham*	1797	−1.27
Morgan, *The Wild Irish*	1806	−1.31
Stowe, *House and Home*	1865	−3.09

CHARACTERIZATION (CONSTRAINT)

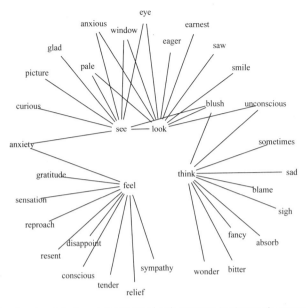

5.5 Collocate network for keywords associated with interiority in nineteenth-century novels with female heroines written by female authors.

on display here). Perception seems far more mixed: *gladness* and *curiosity* are associated with seeing, while *blushing*, *smiling*, *eagerness*, and *earnestness* are associated with looking. Such heterogeneity is worth noting, as is the relatively low interconnectivity of these terms, that each one has relatively distinctive semantic contexts of its own.

Such a representation by no means provides definitive answers, but it can offer suggestions for developing hypotheses and further experiments. It gives us that quantitatively informed interpretive framework that I spoke of in the introduction, a semantic context for understanding what makes these characters' experiences unique to the nineteenth-century novel. They help us situate our observations in a more contextually dependent way, away from our own biases and interests that we inevitably bring to the text, but that are also never entirely left behind even in this numerically informed context. How we read a network or structure an experiment is still a reflection of our own implication in the act of reading, a point to which I will return in my conclusion.

In the remainder of this chapter, I want to focus on a particular quality of interior experience—here indicated by the keywords of *anxiety* and *anxious*, which serve as binding agents between the different actions—and a particular place of visual experience—the window. I

CHAPTER FIVE

want to think about the way women's cognitive and perceptual experience is marked by an unstabilizing intensity in the nineteenth century and what that implies for the type of subjectivity that is being put on display. But I also want to keep an eye on where these experiences are happening, to derive a sense of that scene-act ratio that so fascinated Kenneth Burke in his *Grammar of Motives*. What are the relationships between setting and behavior, and what does this have to tell us about women's interiority bias in the nineteenth-century novel?

Windows and Souls

"Strong anxiety winged the feet of Victoria," writes Charlotte Dacre in *Zofloya* (1806), "and soon she reached the door which had already excited in her mind ideas so various and confused." Here we see something of the agitated female archetype at work: anxious, confused, a mind overly impressed by emotion, and a body standing at the verge of some threshold. This is what it means/takes to cross a boundary as a female character at the end of the eighteenth century. Risk and potentiality are equally mixed. "Yet she looked forward to the future with a trembling anxiety," writes Radcliffe in *The Romance of the Forest* (1791), "that threatened to retard her recovery, and which, when she remembered the words of her generous benefactress, she endeavored to suppress." The anxious heroine is also one for whom futurity looms large. "Lifting up her face Katy looked at her sister with a wistfulness which told how anxiously she waited for the answer," writes Mary Jane Holmes in *The Cameron Pride* (1867). "Her earliest impulse was," writes Maria Cummins in *Mabel Vaughan* (1857), "to possess herself of every particular; her next, to rid the house of strangers; and her last thought seemed to be of the poor sufferer, over whom Mabel hung, in an agony of suspense, while Sabiah wrung her hands, groaned and anxiously asked herself—'how will sister Margaret put up with this new trouble?'"

There is in these moments of intensity a hesitation or prevarication, a moment of doubt that we saw in the last chapter that was one of the novel's core identities. Doubt suggests an uncertain future, an unknowability about what is to come. As it turns out, women in novels written by women are 2.1 times more likely to *look forward* than male protagonists by male authors are. They are also 1.35–1.5 times more likely to *wait* or *await* and have a slight predilection for using more future-oriented words (about 7% more, $p = 2.675\text{e-}11$).[44] At the same time, they are also more strongly associated with the past tense (an increase of about 10%). The future is not straightforwardly next, but aligned with a fictionalized past that presages an alternative world to come. "She saw in his countenance the deep workings of his mind," writes Radcliffe in *The Sicilian Romance*. "She revolved the fate preparing for her, and stood in trembling anxiety to receive her sentence." This is not a world of nostalgia, but one that is deeply concerned with *that which might be*. It shows the extent to which the straightforward Rousseauean stance—the celebration of the prelapsarian moment so common to male Romantic novelists like Novalis who always want to go home—is an articulation of privilege, of a subject for whom the future is far more determined (thus allowing the past to assume greater importance). In the novel of the cogitating heroine, by contrast, the sentence is more often about what is to come.

These results should give us pause to think about differing arguments about female interiority in the nineteenth century. Far from a form of retreat from the world, or even an implied passivity, it is clear from these examples that this cogitative world of thinking and feeling is not only highly agential—there is an activity here—but one that is strongly anticipatory. There is a profound potentiality to subjectivity associated with women that is being articulated through the veil of introversion, one that does not align well with notions of retreat that are often associated with these subject positions. These novels can be read in some sense as a larger social experiment in what it means to prepare the ground for something new, to be open to a thoroughly transformative role in the world.

One of the implications of this stance is the way it comes to be informed by a larger sense of intensity. We often associate interiority with a notion of depth, a deep verticality to being, but when we look more closely at the semantic frames of these types of experiences for nineteenth-century women, we see how words like *strong* and *strength* play a crucial modifying role (not to mention that many of the subsidiary emotions like reproach and resentment are themselves intense). One of the novelists who has the highest predilection for using the vocabulary of strength in this group is George Eliot, someone whom we might not immediately associate with the discourse network of emotional anxiety. It is interesting to see the way the emotional intensity linked to the dual sense of threat and possibility that surrounds the figure of late eighteenth- and early nineteenth-century romance turns in Eliot's work into an altogether different linguistic profile.

When we look more closely at Eliot's work, we see how her overinvestment in the vocabulary of strong feeling is not associated with a sense of openness to the world, however intimidating that world may be for the heroines of romance; more often, it is associated with a sense of nostalgia and negativity. When Eliot uses a form of the word *strong* in her novels, she is 1.6 times more likely to use a negative word, whereas it is the exact opposite for other women writers, who are over three times more likely to use a positive sentiment in this context.[45] Words like *painful, sorrow, fear, agony, hard,* and *petty* appear more often in Eliot's work in proximity with *strong* than other women. As she writes in *The Mill on the Floss*, "No—she must wait—she must pray—the light that had forsaken her would come again: she should feel again what she had felt, when she had fled away, under an inspiration strong enough to conquer agony—to conquer love: she should feel again what she had felt when Lucy stood by her, when Philip's letter had stirred all the fibres that bound her to the calmer past." We see how the word *again* anchors the semantics of feeling in this sentence. Or, as Eliot articulates toward the close of the novel, "Vaguely, Maggie felt this;—in the strong resurgent love towards her brother that swept away all the later impressions of hard, cruel offence and misunderstanding, and left only the deep, underlying, unshakable memories of early union." The vagueness of impression does not presage a transformative future, as in the opening door of Radcliffe. Instead, it signals a resurgent memory of an earlier union, "to feel again," as Maggie traverses the cruelty of misunderstanding that had intervened in the years between.

What we see happening in Eliot is the way she is rewriting the discourse of the anxious heroine, drawing strongly on this vocabulary but situating it within a very different temporal framework. In this, Eliot marks a kind of pivot, placing her at the tail end and beginning of two intersecting phenomena. As she moves away from an earlier form of female subjectivity premised on urgency and confusion, she also ensconces herself within a pending discourse of nostalgia and narration that would emerge as one of the more canonical ways of organizing (male) prestige. She appropriates, we might say, that Rousseauean fantasy of the return and inscribes herself in the process into the literary republic of letters. She merges the cogitative with the restorative and, in the process, generates a powerful template of high cultural storytelling that still resonates today. As I have shown elsewhere, nostalgia is one of the principal ways through which contemporary prizewinning novels differentiate themselves from more market-driven bestselling writing.[46]

If profound feeling was one of the ways through which female subjectivity was implicated in this period, the act of perception interestingly takes us in the opposite direction. If we focus on the figure of the *window* as a privileged space of female visuality (as in fig. 5.5 above, where it conjoins both seeing and looking), that is, if we ask about the ways in which seeing and being seen structure the interactive space of female protagonists in the nineteenth-century novel, we find that one of the window's primary functions is to de-escalate this discourse of intensity.

Windows for female protagonists are first and foremost spaces of retreat. "I saw nothing of Glorvina until evening, except for a moment, when I perceived her lost over a book, as I passed her closet window, which, by the Morocco binding, I knew to be the Letters of the impassioned Heloise," writes Sydney Owenson (Lady Morgan) in *The Wild Irish Girl* (1806). "She was sitting at the end of the dresser," writes Elizabeth Gaskell in *Mary Barton* (1849), "with the little window-blind drawn on one side, in order that she might see the passers-by, in the intervals of reading her Bible, which lay open before her." Reading by the window can either be intense ("lost over a book") or intermittent ("the intervals of reading her Bible"). It is telling that a novel is framed as the genre of immersive reading while the Bible is perceived as an intermittent pastime.

As spaces of reading, windows also function as spaces of pensiveness. They can be sites of resolution, as when Dacre writes in *Zofloya*, "Victoria, after sitting for an hour at the window, with a mind still persevering in the resolution to be firm, sought likewise her bed, and soon forgot the vexations of the day in slumber." Or they can be sites of reverie: "One evening Mary stood leaning against the window, looking earnestly, wistfully upon the beautiful tints which ever linger in the western sky" (Augusta Evans, *Inez* [1855]). But they can also be sites of confinement, confirmations of women's relative lack of social mobility during this period: "She heard the sound of merry parties setting out on excursions, on horseback or in carriages," writes Gaskell in *Ruth* (1853). "And once, stiff and wearied, she stole to the window, and looked out on one side of the blind; but the day looked bright and discordant to her aching, anxious heart."

If windows are where women are often seen to be left behind, they can also be sites of encounter. As Mary Wollstonecraft writes in *Maria, or The Wrongs of Woman* (1798), "Maria was again true to the hour, yet had finished Rousseau, and begun to transcribe some selected passages; unable to quit either the author or the window, before she had a glimpse of the countenance she daily longed to see; and, when seen, it conveyed no distinct idea to her mind where she had seen it before." Putting the book down introduces the face or countenance, one that initiates a state of narrative perplexity. As we see in *Pride and Prejudice* in a similar vein: "But, on the third morning after [Mr. Bingley's] arrival in Hertfordshire,

CHAPTER FIVE

[Mrs. Bennet] saw him from her dressing-room window enter the paddock and ride towards the house. Her daughters were eagerly called to partake of her joy. Jane resolutely kept her place at the table; but Elizabeth, to satisfy her mother, went to the window—she looked—she saw Mr. Darcy with him, and sat down again by her sister." Windows are where this state of recognition takes place, a core space of acknowledgment of social discord. "'La!' replied Kitty, 'it looks just like that man that used to be with him before—Mr. What's-his-name. That tall, proud man.'" There is missing information that the window, the space of perception, allows to be filled in. As Frances Burney writes in *The Wanderer* (1814), "Confounded, speechless, she went to one of the windows, and standing with her back to it, looked at him with an undisguised amazement, that she hoped would lead him to some explanation of his behaviour, that might spare her any serious remonstrance upon its unwelcome singularity." Here the window is not gazed through but is used as a frame for resolution, to move past the space of confoundment and amazement.

In chapter three, I spoke about the significance of *Begegnung* or encounter for women novelists, specifically in Sophie Mereau's rewriting of the topos of life and art in her epistolary novel *Amanda und Eduard*. The immersive liquidity of the male Romantic hero's sensibility before the beautiful object gives way in Mereau to the resistant, confounding encounter with another. Here we see how windows, homes, but also carriages are where female characters experience this sense of "being opposite" from someone or somewhere else, perceptual moments where interiority turns outward.

When we look at the words that are more likely to occur in sentences where windows are present, we find an interesting mix of words to denote stationariness (*standing, leaning, sitting, waiting*, all 2.5–6 times more likely). Things happen "slowly" more often at the window, just as one is more often "still." Stasis becomes a means of experiencing an interior transformation, one that focuses on certain parts of the body (*eye, hand*) as well as certain objects, like *letters*.[47] Windows, it turns out, are spaces of significantly decreased emotional intensity. Sentences with windows exhibit significantly lower levels of both positive and negative emotion when compared to sentences with female protagonists overall.[48] The window is where women go to convene with another at a distance or themselves, to encounter something else (a garden, a street, sky, rain, a carriage, or a resolution, all common words associated with windows). The window marks a moment of turning inward, a sense of, in Dacre's words, "resolution." The strong feeling of female subjectivity that we found out in the wild, as it were—where the female protagonist is herself a kind of medium of intensity for readers—is recuperated around the physical portal of the domestic window, made more resolute. Interiority takes on different meanings depending on where one is situated.

The Afterlife of Interiority

When critics talk about being "hailed by a text" (D. A. Miller) or "too-close reading" (Frances Ferguson) and then use Austen's *Emma* as an example of these experiences, we can see how a specific historical constellation connected with the practice of characterization is being generalized into a more universal mode of reading.[49] Austen's *Emma* is not an accidental choice through which to register feelings of depth and attachment. As we have seen, it is in many ways *the* archetypal example of such immersive or invitational reading. The cogitative investments of Austen overall and *Emma* in particular are the ideal scenes upon which fantasies of readerly identification—of being hailed or being too close—are able to transpire.

But in generalizing a historical moment outward, critics like Miller and Ferguson overwrite the specifically gendered aspect of this story, as well as the specifics of that gendering—the various ways this type of characterization serves particular social interests at a distinct time and place. With their focus on deep, cogitating, and inviting female minds that transcend time, they overlook the qualities of intensity and temporal futurity that were some of the core features of female interiority of the period and that have distinct political overtones. In their nostalgia for a certain type of character, they miss the forwardness of these characters' own identities and the cultural work it was meant to perform.

At the same time, when we universalize these character types, we also miss the particular gender politics that could surround specific types of introverted experiences. Observational spaces like the window were not only places of respite from the emotional intensities of such social thresholds—of women's sense of what is to come. They were also places to escape the impositions of male observation. Women have a much greater chance of *not* being in the presence of men near windows than they do in general (about 51% more likely, $p = 6.712\text{e-}12$).[50] While there are plenty of examples of agitated women at the window, there are far more where a greater degree of equilibrium is being found, an equilibrium that is accentuated by the absence of a male character. Alongside the history of madwomen in attics or anxious and trembling heroines, there is another story to tell about the emotional equilibrium, the sense of *resolution*, that is achieved by women in novels through spaces of perception. Interiority by the window becomes a means of resistance, the resistance to always being seen.

By essentializing this type of behavior as either universal or only feminine, we also miss a crucial way in which this mode gradually evolves into new ends. As Ted Underwood has shown, female characters become increasingly less semantically distinct over the course of the novel's development.[51] The unique profile of the cogitating-perceiving heroine will eventually fade, but not the identity of the profile as such. Continuing to replay the love affair with Jane Austen and close reading, in other words, not only occludes why such characterization was happening back then. It also occludes the ways in which such characterization continues to play a role in the novel today, albeit in a very different social and aesthetic context. Putting Austen or gender before characterization overlooks the continuity of a certain way of characterizing, of how this sense of constraint can serve different social purposes at different points in history. It is precisely these kinds of insights that computation is well suited to understand.

I want to conclude with a look at where this cogitative-perceptual bias of female characters goes in the twentieth and twenty-first centuries. In the last chapter, we saw how much of what comes to indicate fictional discourse in the novel becomes stable over the longer duration of time. If we compare our cogitative and perceptual vocabularies within nineteenth-century novels and those from the twentieth, we see a slight increase of about 1% on average per novel (roughly about five extra words per page), but with a much greater amount of variance in the twentieth century (meaning a greater spread between high- and low-scoring novels).[52]

What changes even more significantly is that now a majority of authors are male (62%) for characters who register such behavior to a significantly high degree (defined as the 95th percentile and above), whereas before it was women who were the majority, at 57%. Second, we also find that 70% of these novels have male protagonists, where for the nineteenth century no more than 48% of highly cogitative-perceptual protagonists were men. Taken together, this means that there is a 230% increase in the likelihood in the twentieth century that a highly introverted main character will be a man written by a male author, compared with the nineteenth century.

Table 5.6 The genres of twentieth-century novels with main characters that exhibit levels of introverted behavior in the 90th percentile or above. Values refer to the percentage of novels represented by a given genre in the 90th percentile or above.

Genre	%
SciFi	38.1
Mystery	24.3
Literary fiction	13.4
War	11.4
Romance	6.4
Historical fiction	4.9

At the same time, the types of novels that this behavior is associated with are also disproportionately oriented around science fiction and fantasy, with a significant portion of literary fiction mixed in as well as mysteries and war stories (table 5.6). This is a world of characters by Zane Grey, C. S. Lewis, Ernest Hemingway, Somerset Maugham, P. D. James, and Michael Crichton.

This trend can be confirmed when we look at the contemporary collection of well-selling writing. An analysis of variance test tells us that romance and young adult protagonists are abnormally low in interiority, while it is only the main characters of science fiction who are abnormally high (meaning prizewinners, bestsellers, and mysteries show no significant difference from one another).[53] Overall, we see a 1.9% increase in the median value of cogitative-perceptual vocabulary in science fiction compared to other genres, not including Romance and Young Adult, which translates into roughly 9–10 more words of this type *per page*.

When we plot a similar semantic network for our four key behavioral terms for contemporary science fiction, we see how the topos of interiority has strongly shifted (fig. 5.6). The sense of emotional intensity associated with characters' "thinking" has moved away from notions of *bitterness* and *sadness* (perhaps more generally with the idea of regret) toward the more immediate sense of *danger* or a *problem*. "Feeling" has shifted as well from a more interpersonal spectrum that ranged in the nineteenth-century novel from *gratitude* to *resentment* toward a framework in the twentieth century driven by embodiment (*panic, pulse, pressure, muscle, gut*). Sight, too, shifts from the more framed and potentially contemplative observations of *pictures* or *windows* in the nineteenth century to the more recent perspectival attention to *angles, shapes*, and *distances*.

Far from being solely about the hardware of the future, seen from

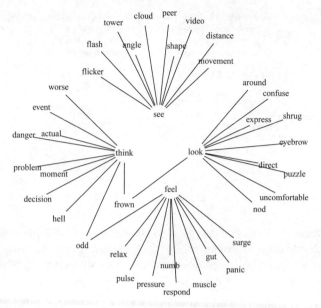

5.6 Collocate network for keywords associated with interiority in contemporary science fiction novels.

this vantage point science fiction is a genre concerned with thinking, sensing, and feeling the uncertainty of human embodiment. It wrestles with the fate of our bodies in an increasingly technologized world. A sense of urgency remains constant, only now instead of encounter it is about self preservation, the continuity of corporality. This is yet another constraint of character, one we inhabit more urgently every day.

SIX

Corpus (Vulnerability)

"To be in any form, what is that?" WALT WHITMAN

What does it mean to be vulnerable, to be open to the world? And how might we communicate this vulnerability in writing? Since Cicero, we have talked about the writer's "corpus" to refer to his or her writing as a body of works.[1] The corpus is meant to be organic and integral—well connected—but also distinct and whole. The corpus marks limits. It has a beginning and an end. It is the material complement to the life of the author. The one gives form and meaning to the inchoate nature of the other. Not for nothing was the final volume of Rousseau's collected works a report on the autopsy of the dead writer's body.[2] Vulnerability marks the end of the corpus, in a scriptural as well as biological sense.

This chapter is about studying what it means to imagine writing as a body, as something that has a distinct shape or form, but is also subject to being vulnerable. I am interested less in what makes a writer "unique," and more in trying to understand those moments when a writer opens his- or herself up to change, novelty, or difference. How radical or gradual are these movements? How permanent or fleeting or even recurrent? Is there something called "late style," a distinctive signature that characterizes the end of a career, as the contours of the aging body are mapped onto the weave of writing? When and how do we intellectually and creatively exfoliate?

Much of the early success surrounding computational literary criticism pertained to the category of authorship.[3]

CHAPTER SIX

Detecting whether an unknown work belonged to a given writer proved to be a task that computation was very good at solving. One of the surprising insights of this work has been that the stylistic signatures of writers are not only detectable, but are often detectable through the least meaningful types of words.[4] The writer's subjectivity comes across most recognizably through the emptiest of signs. This is yet another way of thinking about the "author function."

In this chapter, I want to explore a different approach to authorship, not as something static and distinct, nor as a juridical necessity, but as a process of change and evolution. While we have very successful ways of detecting "authors" or "style," we have considerably fewer techniques for talking about change, the nature of the variability within an author's corporal outline.[5] And yet such change—and the assumption of its very existence—is built into the fabric of how we think about writers' lives and the life of writing. We talk about the evolution of Walt Whitman, the conversion of T. S. Eliot, or the early or classical or late Goethe. In each case, we are establishing relationships between different parts of a writer's works and arguing for their particular shape or organization.

Such judgments are not entirely out of sync with the use of traditional methods. It is possible to read all of the poems written by a single poet or all of the novels written by a novelist (though all *works* by a prolific author already begin to stretch the bounds of plausibility—Goethe's collected works run to 148 volumes and do not include his voluminous archival material). More challenging, however, is the ability to make informed judgments about the overall configuration of relationships on display. The number of connections within even a single corpus quickly becomes immense. More challenging still is the ability to make comparative judgments *between* authors, to understand the particular ways in which the shape of one career might differ from another. Computation can give us the means of making these kinds of comparisons, of creating stable frameworks across different collections and even different languages. It can help us identify the array of figures that a body of writing can take.

Consider for example the work of Walt Whitman. Over the course of his lifetime, he wrote, rewrote, and expanded a single collection of poems that has since become an American classic, *Leaves of Grass*. We can, as traditional criticism has done, test the extent to which the various editions differ from one another and how those differences transform over time. Looking at the lexical, semantic, syntactic, and phonetic similarities between the editions using a method known as hierarchical clustering, we can see how these relationships largely correspond to the re-

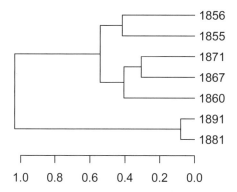

6.1 Stylistic affinities between the different editions of Walt Whitman's *Leaves of Grass*.

ceived narrative of Whitman's life (fig. 6.1).[6] If we look initially at the document pairs that are generated, we see how the early editions cluster together, as do the late editions, just as we can observe a local division between the pre– and post–Civil War editions. And yet, if we move up our tree (or in this case further left) to observe the larger divisions that exist between the editions, we do not see a clear binary on display between the pre– and post–Civil War editions—one of the core convictions of scholarship on Whitman. The computational model's tripartite division between early, middle, and late editions puts pressure on the scholarly consensus about the war as a key dividing point in Whitman's career. Indeed, if we were to divide the editions into two groups, the dendrogram suggests that it would cleave not around the war but around the more biologically inflected category of lateness. There is something about those later editions that pulls them away from the rest, according to the logic of the model being used here.[7]

Even as these somewhat provocative distinctions emerge from the computational model, we are still not far from a framework indebted to traditional criticism. The retroactive assessment of the partitions that comprise a writer's career has been and remains one of the more ingrained approaches to studying authors. Instead, we can try to model this development as a dynamic process, where we do not have access to knowledge of the whole, but only the output up to a given point in time. In this model, we would be representing something closer to the author's own experience (where as a poet I only know what I have done *so far*, not where I will end up on my deathbed).

In the case of Whitman, we could think of each edition in these same terms, where the book becomes a microcosm of the career. As the edition evolves, we can make judgments about the relationships between pages

CHAPTER SIX

6.2   Network of Walt Whitman's 1891 edition of *Leaves of Grass*. Each node is a page of the edition. Links are drawn between pages that are the most similar to each other. The nodes darken as the page numbers increase.

over the time of the edition, where time would stand in this case for something akin to the reader's, rather than the writer's, experience. As a reader of *Leaves of Grass*, I do not yet have knowledge of the entire book, but only that which I have already read (a point that may be true even for re-readers). Such a model would allow us to see the extent to which the edition moves away from its earlier sections, curls back on them, or rhythmically (or happenstancely) bounces back and forth between them. It would allow us a new kind of insight into the different natures of the editions beyond the stylistic affinities between them.

Figure 6.2 provides a portrait of what such a model would look like. Here, the nodes of the network represent standardized pages of a particular edition (in this case the final 1891 edition), and the edges represent the degree of stylistic similarity between them.[8] A page can only connect with a page prior to it (not after) and only selects the closest page to it in terms of stylistic similarity unless there is an ambiguity of choices, in which case it selects all pages until it hits a significant drop in similarity (keeping usually no more than 2–3 pages). While the network itself may not convey much information to our naked eyes, when we measure certain features about each edition, we begin to see an interesting story emerge concerning the evolution of Whitman's work.

The graphs in figure 6.3 provide a snapshot of different measures we can use to understand these temporal networks of the editions of *Leaves of Grass*, that is, how the editions evolve both internally and from edition to edition. The first three (a–c) provide a sense of how deeply the work reaches back into itself in its construction in terms of stylistic similarity between pages. In (a), we see the average distance between connected pages in terms of how many pages apart they are (these numbers have been normalized as percentages to control for the different edition lengths). So, in the 1860 edition, the distance between connected pages represents on average about 15% of the total work, while by 1891 that number rises to closer to 20% as connected pages are drawn from further reaches of the volume. The second measure (b) allows us to see what percentage of the most similar pages are less than five pages apart, while (c) looks at the percentage of most similar pages that are more than twenty pages apart. As we move forward in time, we can see how in: (a) each page's most similar page tends to become more distant relative to the overall length of the work; (b) the percentage of pages that refer to the most recent five pages before them tends to drop; and (c) the percentage of pages that refer to a page more than twenty pages away tends to increase. These measures suggest the way each edition, even when we control for their changing lengths, tends to reach deeper into itself as Whitman ages.[9]

The next measure (d) looks at the standard deviation in terms of the stylistic similarity between pages for a given edition. The higher the standard deviation, the more variable the similarities are between connected pages. The declining standard deviation over time suggests that Whitman's vocabulary is getting more standardized as he ages and revises (with the exception of the final edition).

Finally, the "vulnerability score" (e) tells us the percentage of pages needed to be removed before the network breaks into two components of relatively similar size (where pages are removed based on their degree of connectivity, starting with the most highly connected pages).[10] The idea here is to model what it would be like if a single poet were actually two different people—how many deletions of one's career are required

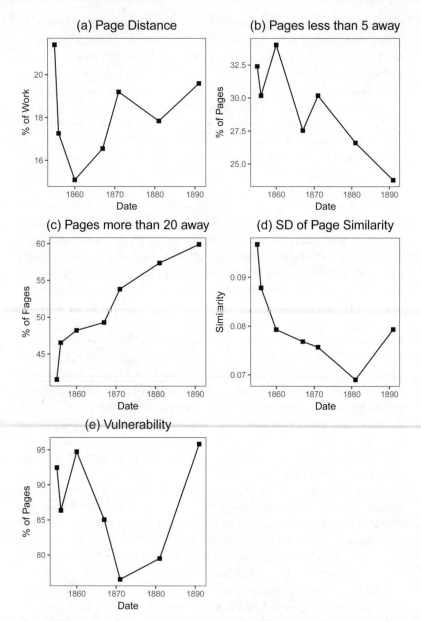

6.3 Network measures to assess the interwovenness of the different editions of *Leaves of Grass*.

before one looks like a double of him- or herself? The fewer pages needed to be removed, the more fragile or vulnerable the network (or work or career). For example, in the 1871 edition, the least vulnerable of Whitman's works, over 20% of nodes need to be removed before we see two distinct components of relatively similar size emerge, whereas it only takes the removal of just over 5% of pages (or about 18 pages) in the 1891 or "deathbed" edition to arrive at the same point. It suggests that Whitman's editions are becoming more tightly interwoven over time—they are harder to break apart into coherent units—with the key exception of the final deathbed edition.

Taken together, then, we see how Whitman's editions appear to be doing two things at once—they become more expansive over time in the way they refer to themselves, but they also become less heterogeneous, that is, less vulnerable. The exception to this rule is the final deathbed edition, which interestingly is often considered a variant of the 1881 edition and not a true "edition" at all. But these measures suggest that the additions Whitman made have an important impact on the overall structure of the work as it relates to itself, giving it a unique signature within an otherwise fairly steady developmental process. That final edition begins to open itself up to new kinds of expression and new kinds of connections. It becomes more vulnerable, like the aging poet who compiled it.

In what follows, I will be using these and other measures to study the shape of writers' careers. To do so, I will be working with a trilingual collection of about 30,000 poems written in French, German, and English by 78 poets whose work spans the past three centuries (POETRY_CW). The data used here is not to be taken as authoritative critical editions in digital form (though much of the data is indeed drawn from critical editions). We are still a long way from that. But in compiling appropriate representations of a diverse range of poets—Victor Hugo, J. W. Goethe, William Wordsworth, Christina Rossetti, Adrienne Rich, Nicole Brossard, Amiri Baraka, Wanda Coleman, Seamus Heaney, Else Lasker-Schüler, Paul Celan, and Francis Ponge, among many others—we can begin to make comparative assessments about the nature of the poet's corpus: its shape, its change, and its vulnerability. What does a career look like?

Whitman is of course not an arbitrary choice to begin with. As commentators have repeatedly pointed out, few poets have been more obsessed than Whitman with thinking about the body of poetry, the implications of *embodiment* for creative writing. Computation can help make this fact more visible, not as an intact, heroic whole, but as an exercise in identifying the poet's vulnerabilities, those places where he or she opens him- or herself up to something as yet un- or under-expressed.

CHAPTER SIX

This chapter is an attempt to establish some of the groundwork for understanding that larger relationship between corporality and expression.

On Similarity

At the basis of the idea of the "corpus" is a belief about similarity—these poems are more similar to each other than others. They belong *here*. There may be groups within groups—such as the notion of authorial periods that I will discuss later—but overall the corpus is held together by a sense of affinity, of things belonging together. But what does it mean for two poems to be "similar"?

Before proceeding to the representation of the entire corpus, I need some way of making explicit this more local problem of similarity between individual poems. What features matter? How will they be treated? And what counts as similarity? For the purposes of this chapter, I will define similarity in the following way: *two poems are similar if they share a significant number of features across four key poetic dimensions that include the lexical, semantic, syntactic, and phonetic aspects of a poem*. While there are many more dimensions of poems that one may wish to understand, these four encompass a diverse representation of the foundations of poetic meaning. I define the notion of "sharing" features as the cosine similarity between these feature vectors.

At the same time, my model does not treat all dimensions equally. Instead, I prioritize semantic similarity because of the way it aims to capture the conceptual drift of two poems rather than rely exclusively on their lexical intersection. Semantic similarity is derived here using a process known as latent semantic analysis (LSA), which transforms the word distributions of a given corpus into more general "semantic" distributions using the process of single-value decomposition.[11] A poem may use different words, such as *tree* or *vine*, to express similarly related thoughts. This similarity will in theory be captured by the overlapping words associated with these two words, which LSA is designed to capture. As we will see, the associations made through LSA are, in the case of poetry, quite powerful and often moving. Occasionally they are too associative, like an overeager student who wants to see connections where none exist. The correction via other, more literal, features like syntax, lexeme, or phoneme helps minimize those leaps and grounds the associations in the language, syntax, and sound of the poems. A full discussion of the model can be found in the appendix. Here I want to provide a brief illustration of the different kinds of choices this model makes when compared to a more literal one that takes only word features into account.¹²

Let me take as an example Muriel Rukeyser's highly anthologized poem "Summer, the Sacramento," which appears in her collection *The Green Wave* (1948). It is a beautiful poem attuned to the shifting colors of the Sacramento River, near the foothills of Mount Shasta.

SUMMER, THE SACRAMENTO
To this bridge the pale river and flickers away in images of blue.
And is gone. While behind me the stone mountains
stand brown with blue lights; at my right shoulder standing
Shasta, in summer standing, blue with her white lights
near a twilight summer moon, whiter than snow
where the light of evening changes among these legends.

Under me islands lie green, planted with green feathers,
green growing, shadowy grown, gathering streams of the green trees.
A hundred streams full of shadows and your upland source
pulled past sun-islands, green in this light as grace,
risen from your sun-mountains where your voices go
returning to water and music is your face.

Flows to the flower-haunted sea, naming and singing, under my eyes
coursing, the day of the world. And the time of my spirit streams
before me, slow autumn colors, the cars of a long train;
earth-red, earth-orange, leaf, rust, twilight of earth
stream past the evening river and over into the dark of north,
stream slow like wishes continuing toward those snows.

"Summer, the Sacramento" is a poem about change and flow, what is gone, but also what runs against the current. It abounds in color, but also sound, in order to think about how we orient our sense of self in the world ("the time of my spirit"). It begins by observing a "bridge," and concludes with "wishes continuing toward those snows," snows that are located not in the direction of the river's flow, but in the opposite direction, "over into the dark of north." It uses synesthesia ("music is your face"), but also alliteration ("green growing, shadowy grown, gathering streams of the green trees"), within an overall shifting palette of greens and blues that turn by the poem's end to reds and rust. It stands in a long tradition of poetic thinking in which the river and the mountain stand for the twin poles of time and timelessness.

CHAPTER SIX

If we ask what poems in Rukeyser's career are most similar to it, the answers we are given will depend heavily on the features we consider. There is of course no overall, value-neutral notion of "similarity." By way of illustration, I want to present two examples, the first a poem that is judged to be similar to "Summer" when we consider only words (the lexical model) and the second a poem considered to be similar when we use the combined model discussed above.

First, a poem that the expanded model rejects but that is selected by the lexical-only model (where the words that overlap between the two poems are in bold):

A RED BRIDGE FASTENING THIS CITY TO THE FOREST
A **red bridge** fastening this city to the forest,
Telling relationship in a stroke of steel;
Cloud-hung **among** the mist it speaks the real,
In the morning of need asserts the purest
Of our connections: for the opposites
To call direct, to be the word that goes,
Glowing from fires of thought to thought's dense **snows**,
Growing among the treason and the threats.
Between the **summer** strung and the young city,
Linking the **stone**fall to the **tree**fall slope,
Beyond the old **namings** of body and mind
A **red bridge** building a new-made identity:
Communion of love opened to cross and find
Self the enemy, this moment and our hope.

The most obvious connections at work here are how both poems share the figure of the bridge as well as that of snow ("thought's dense snows"). The bridge, like the river, connects; it makes possible some cognitive experience, one that is largely categorized by opposites: "body and mind," "stonefall to the treefall slope." Here the bridge is about the act of naming ("beyond the old namings"), of finding the right new word to encompass these contrasts. "Summer, the Sacramento," however, is more about a process, the act of looking up- and downriver, and the way moving upriver creates a new kind of stream into something unknown ("like wishes continuing toward those snows"). The vibrant colors, alliterations, and synesthesia are all missing in the red bridge poem, just as the sense of "treason" and "threats" in the red bridge is missing from the lying tranquility of "Summer." In many ways, it is the urban/nature divide that divides these poems; whereas "Summer" is solely concerned

CORPUS (VULNERABILITY)

with the knowledge of nature, "Red Bridge" wants to think about how to merge steel, stone, tree, and mind into a more unified whole.

A second example shows us a poem that was selected by the expanded model, but rejected by the lexical-only model:

BLUE SPRUCE
Of all **green** trees, I love a never**green**
blue among dark blue, these almost black
needles guarded the door there was, years
before the **white** guardians over Sète
. . . that's **Sea** France at the **Sea** Cemetery
near Spain where Valéry . . .
those short square Mediterranean
man and woman
couple at the black-cut **shadow** door
within the immense marine
glare of noon,
and on the beach
leaning from one strong hip
a bearded Poseidon
looking along the surface of the **sea**
father and husband there he **stands**
and an invisible woman him beside
blue-eyed **blue**-haired **blue-shadowed**
under the sun and the **moon**
they blaze upon us
and we waiting waiting
swim to the **source**
very **blue evening** now deepening
needles of **light** ever new
a **tree** of **light** and a **tree** of darkness
blue spruce

"Blue Spruce" is from Rukeyser's final collection of poems, *The Gates* (1976), which tends to be more abstract and associative than her earlier work, composed of diverse visual fragments as well as much greater amounts of syntactic irregularity. "Blue Spruce" begins with a reflection on the tree of its title ("Of all green trees, I love a nevergreen blue among dark blue"), and we can see the color palette of "Summer" repeating itself here as well as the alliterative, repetitive act of description. The poem moves quickly to reflections on the "Sea Cemetery" at Sète, where

CHAPTER SIX

Valéry is buried (and also the title of a long poem by him). The primary body of water has shifted from the stream to the sea, but the emphasis is still on the effect of observing the liquid surface, or the liquidity of surface. "Blue Spruce" closes with the statue of Poseidon, "looking along the surface of the sea." In the process of personifying the statue, we adopt his point of view.

What we see happening in this poem is the way Rukeyser rewrites Valéry's lyric engagement with the sea as one of metempsychotic personification, where the poet (and reader) occupy the point of view of an enlivened object. "We" replaces Valéry's "I," as well as her own "I" from the earlier poem, "Summer, the Sacramento." Despite these divergences, "Blue Spruce" and "Summer" are both ultimately concerned with a source and its attainment, with reflections on the basis of language. Where the former moves toward the dark north amid the contrasting light of mountain and river, here we end in the spruce's shadows, "needles of light ever new / a tree of light and a tree of darkness." The spruce, and its blue-nevergreenness, is what makes possible this vision born from light and dark that we saw at work in the foothills of Mount Shasta on the banks of the Sacramento.

While the expanded model reflects some of the literal connections that exist between these two poems (*blue, green, source, tree, shadow, light, white, sea*), it also seems to capture the larger conceptual scope that pervades them. It pinpoints that striving toward a source that is common to both poems, a processual desire for finding the foundations of thought and language through poetry. Not all readers will agree that these poems belong together or that these commonalities outweigh those of "Red Bridge." Some might also prefer other poems to this one. A great deal more research is needed to understand the ways in which readers make connections between poems and how we might potentially model these associations.

As I will show in the next section, we can also design models that do not make hard and fast choices, but try to understand the larger distribution of similarity in which a given poem operates. In either case, I would argue that the multidimensional model used here offers a reasonable way of capturing both the linguistic and syntactic commonalities between poems as well as something less tangible, yet no less important. Whereas the connections of "Red Bridge" to "Summer" felt thin (a mere two bridges), those of "Blue Spruce" to "Summer" feel thicker, deeper, but also somewhat more elusive. Ultimately, if we want to model connections between poems, it is these more evanescent associations that we want to identify or make visible, even if there will be plenty of examples that feel less plausible. Error and insight are never far apart.

Vulnerability

If similarity is a measure of the ways in which poems share a certain conceptual and linguistic intersection, then vulnerability is the moment when this overlap begins to falter. Vulnerability is when a poem, or group of poems, lacks a stable framework within the poet's corpus, when it is not well contained by the expectations established by the poet up to that point in his or her career.

In what follows, I will be using two notions of poetic vulnerability. The first is what we might call "global vulnerability," when a poet's overall connective tissue is more easily fissurable, that is, when the network of poems established by a poet breaks down. Vulnerability is traditionally used in network science to measure how well connected a graph is and therefore how vulnerable it is to attack or breakdown. It consists of removing nodes (a process called *percolation*) in descending order of connectivity (so the most-connected node is removed first, then the second, etc.) until the graph is no longer fully connected (meaning at least one node is no longer connected to any other node).[13] In my model, I use as a breakpoint the moment when there are two components remaining where the smaller is at least 50% as large as the larger component. The idea is not to measure the disconnection of any single node, but rather how long it takes before the poet's corpus divides itself into two relatively coherent entities, that is, until s/he becomes double. This is what I explored in the beginning of the chapter with respect to Whitman's final deathbed edition, where very few pages needed to be removed before the edition cleaved into two parts of relatively equal size. Poets who have high global vulnerability scores are poets whose collected works are easily divisible. Take away just a few poems for this group, between 5 and 7% of the total amount of poems (roughly 10 to 25), and their corporal networks begin to break in two.[14]

What this looks like in practice can be seen in figure 6.4. Here we see the network of Annette von Droste-Hülshoff's corpus after the top 17 most-connected nodes have been removed. While many poems float around in unconnected space, we can observe two large components that are relatively near each other in size. These components represent two different strains (even worlds) in Droste-Hülshoff's career. On the

CHAPTER SIX

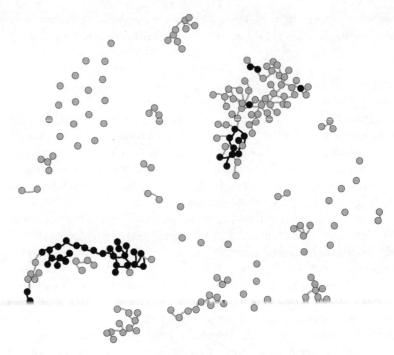

6.4 Network of Annette von Droste-Hülshoff's poetry after removing the top seventeen most-connected nodes/poems. Poems in black belong to the genre of occasional poems written for religious holidays.

one hand, we have the dedicatory occasional poems written for religious holidays on the left (the nodes marked in black). And on the other, her more natural poems in the larger component on the right, which she often titled "Heidebilder," named after the moorlands where she lived. The distinction contains another layer, as "Heide" can also, as it does in English, refer to a heathen ("die Heide" means heath or moor, while "der Heide" is the masculine form of *heathen*). We see Droste-Hülshoff's poems literally cleaving between pagan and Christian semantics, with the latter group denoted by an emphasis on a vocabulary of *God*, *Sunday*, and *becoming*, while the former is captured by objects like the *earth*, *moon*, *sun*, and *air*, along with senses like *listening* and *seeing*. These networks represent a microcosm of a larger nineteenth-century cultural imaginary, one that is visually ensconced within the author's poetic corpus.

The second measure I will be using here is "local vulnerability," which corresponds to those moments in a poet's corpus that depart from previous stylistic expectations (where style is defined as the four-part model described in the previous section).[15] This is when we see a writer moving away from the norms of his or her own oeuvre. One way to do this is to observe the average similarity of each poem by a given writer to all previous poems by that writer. Rather than choose a single best fit for each poem's closest association, as in the global model, we can observe instead the larger distribution of similarity between existing poems and every new poem. Given what the poet has created *so far*, how does the new poem relate to the rest of the work? This allows us to assess not a single, global sense of the poet's corpus, but its more linear change over time.

In figure 6.5, we see a rolling average of the similarity between a given poem and all previous poems across a ten-poem window from the work of four poets. The upper and lower bands represent significance thresholds,

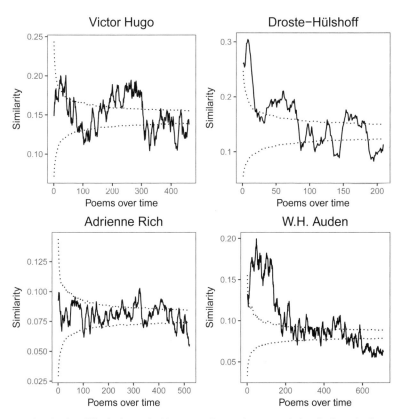

6.5 Local vulnerability in the work of four poets. Poems that appear below the lower horizontal line are significantly different from the stylistic expectations up to that point in the poet's career.

that is, when the degree of similarity or difference between poems is greater than one might find by chance.[16] This is another way of saying that a poem or feature is distinctive. These graphs give us insights into the extent to which a given range of poems departs from the expected norms established by a poet throughout his or her career. Peaks represent moments when poems begin to look more like what one has read, and valleys represent moments when poems fall below the stylistic variation one might otherwise find by chance. They are signs that a poem has exceeded the horizon of expectation established by the poet. "Local vulnerability" is thus understood as the percentage of poems that fall below the lower threshold (the valleys).

While I return later in this chapter to the ways in which local vulnerability can be read in particular instances of poets' careers, I first want to represent the relationship between these notions of global and local vulnerability in order to capture a more holistic picture of poets' careers (fig. 6.6). Poets like Felicia Hemans and Else Lasker-Schüler have collec-

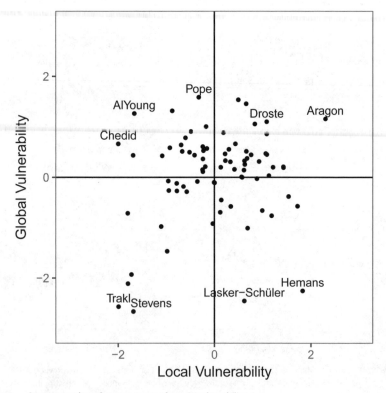

6.6 Career types based on two types of poetic vulnerability.

tions that are difficult to divide in half (low global vulnerability) but also experience a great deal of local experimentation (high local vulnerability). Poets like Al Young or Andrée Chedid, by contrast, seem to have fewer moments of divergence from stylistic expectations (low local) and yet still have globally fragile collections (high global). Annette von Droste-Hülshoff and Louis Aragon are marked by a great deal of both local swings and global fragility (high/high), while poets like Wallace Stevens and Georg Trakl never quite exceed their own expectations (low/low). It is interesting that poets who are known for being more stylistically experimental often seem to have the least varying of careers. Their difference from the world closes them off from their own self-differentiation. A poet like Felicia Hemans, on the other hand, has a great deal of local vulnerability, but those cracks never add up to any sizable global distinctions within her work. The connections throughout her writing are far more marbled than tectonic.

The Poet's Periods

When combined, the vulnerability measures give us some insight into thinking about the shape of a poet's career, in particular as it relates to questions of change and variation. The combination of these features allows us to think about new ways of understanding a poet's work as an entirety that evolves over time and to group poets together by new kinds of criteria. One measure takes a more holistic view of the corpus, while the other looks more linearly at changes over time.

A natural question that arises is whether these windows of vulnerability last long enough to be considered a movement unto themselves, or what we would more commonly call a "period" within an artist's life. Is the signal of change strong enough and does it last long enough to constitute a "breakpoint" in the stylistic sensibility of the poet? And what would we mean by "enough"?

Periods are artificial ways of grouping works of art in a linear fashion, a means of reducing the overall complexity of a corpus into smaller units that are fundamentally time based. Traditionally we have worked with models that have strong biological bases—early, middle, and late periods, for example—but we often consider stylistic variations to be their own periods as well (Picasso's "Blue Period" perhaps being the most famous). In other words, periods are nothing more than clustering algorithms where the rules or instructions for how to divide a corpus into smaller units are implicit or latent, rather than explicit. Computation

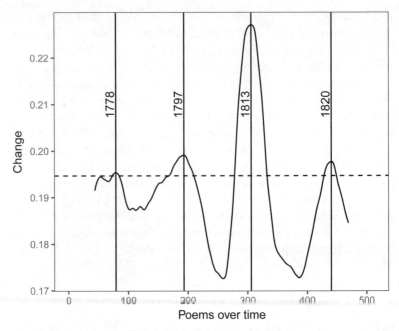

6.7 Predicted periods in J. W. Goethe's collected poems.

allows us to make the criteria through which a period is constructed not only explicit, but also common across all poets.

One way of assessing periodization in a poet's corpus is to use a measure developed in the field of musical stylistics called "Foote novelty."[17] Foote novelty, named after Jonathan Foote, looks for windows of stylistic difference using the same underlying data from the local vulnerability scores. Instead of measuring the entire distribution of relationships between a given poem and all those preceding it, it looks at a local window before and after a poem to assess whether the degree of change surrounding that window is significantly greater than one might find by chance alone (beyond a normal degree of variation).[18] Foote novelty might be thought of as a more generalized version of local vulnerability, one that slides across the corpus over time looking for more aggregated levels of change rather than individual variations.

Figure 6.7 presents an image of Goethe's collected works divided into predicted periods based on Foote novelty, using a window of 20 poems (i.e., +/−10). Here the peaks represent maximum moments of change, captured with the vertical lines, while the dotted horizontal line represents the significance threshold, that is, levels above which we no longer believe are simply due to random fluctuation with the data. All moments above this amount suggest that unlikely degrees of stylistic change are occurring in the corpus.

According to Foote novelty, then, Goethe experiences five distinct periods during his career, the most of any poet in the collection and an amount shared by only two others (Droste-Hülshoff and Émile Verhaeren). The first is the least dramatic and occurs around the poems "Der Fischer" and "Harzreise im Winter," written in the late 1770s, after which we find a period of relative stability that focuses on a great deal of poems aligned with classical antiquity (which Goethe labeled "Antiker Form sich nährend"). The second and more significant turn occurs around the rising importance of the genre of the elegy in the late 1780s and early 1790s. While this period pivots on the elegy "Euphrosyne," the rise in novelty is driven in large measure by the introduction of the Roman Elegies a few years earlier that correspond with Goethe's Italian journey. The next two periods occur in Goethe's so-called late period, the first around 1813, with the clustering together of short epigrammatic poems and what he calls "Gesellige Lieder," while the second occurs in 1820 following the *West-East Divan* poems and the more mystical *Gott und Welt* collection.

In other words, Foote novelty approximates reasonably well much of the consensus surrounding Goethe's oeuvre, while adding some specificity and nuance to that portrait. The 1770s appear as a coherent epoch in Goethe's creativity that begins to shift in the 1780s with a turn toward not only more classical forms, but also ballads and the lyrical poetry that would populate the first Wilhelm Meister novel. The second period corresponds with the major biographical move to Italy and his return that is then articulated through the Roman Elegies, while the third disjuncture puts an interesting spin on how and when to divide the later period. Usually we date it around Schiller's death in 1806, but Foote novelty wants us to see how the epigrammatic poems mark a space of real departure from the rest of the corpus—we know this from the scholarship, but only by negation. These are works that are rarely if ever discussed. Finally, the measure picks up the very late shift to world literature that occurs through the *Divan* poems as well as the more cosmological *Gott und Welt* poems, incidentally some of my favorites.

If we run this model on all the poets in our collection, we can see some general trends begin to emerge (table 6.1).[19] Five periods appears to be the maximum amount in a single career, while some poets are estimated to have undergone no major stylistic shifts at all in their careers (close to one-quarter of all poets). The average age when a poet undergoes his or her first period is estimated to be 28, with the median age being 29. The

CHAPTER SIX

Table 6.1 Summary statistics on periods in poets' careers

Maximum number of periods	5
Poets without a significant period	23%
Avg. age at the end of the first period	28 years
Avg. length of the first period	12 years
Avg. length of the longest period	21 years
Poets with the shortest period at end of life	34%
Poets with majority vulnerable poems at end of life	42%

twenties appears to be the most common decade in which a poet undergoes his or her first major shift, far earlier than I might have thought. A poet's first period lasts 12 years on average, though the median is closer to 8. The range is anywhere between 2 and 56 years. The longest period in most poets' careers is about 21 years long (+/–3 years). About 34% of poets who have more than one period have their shortest period at the end of their lives, suggesting that this shift occurs relatively close to their death. Late periods, however, are not significantly longer or shorter than early periods, meaning there does not appear to be any bias toward swifter change in poets' early years or a particularly late swerve just prior to death. Finally, about 40% of poets have a majority of their vulnerable poems occurring in the final quarter of their lives, meaning that most of their vulnerability occurs relatively late in their lives.

According to the model, periods occur around different kinds of turning points in authors' lives. They are most often predicted to occur around volumes of poetry, where the move to a new book signals a distinct stylistic difference (as in the case of Goethe's *Divan*). Sometimes this can happen very quickly, as in the case of the Belgian modernist poet Émile Verhaeren, who shifts strongly between his first nationalist collection *Les Flamandes* (1883) and his second, more religiously inflected, *Les Moines* (1886). Similarly, three years into her career as a published writer, Muriel Rukeyser will move into a distinctively new style with her second book of poems, *U.S. 1* (1938), though this shift looks considerably less drastic when compared with her postwar transformation in *Body of Waking* (1957). It suggests how the artifact of the book continues to play a very strong role in how poets evolve in their writing.

The pace of change varies a great deal too. While Rukeyser changes quite quickly, it would take Elizabeth Barrett Browning and Victor Hugo until their mid-fifties to undergo their first major shifts. For Hugo, it was precipitated by the experience of exile and took the form of moving to longer narrative poems like *La Légende des siècles* (1859) (though there were earlier variations of note a quarter century earlier, for example around the move from *Les Orientales* to *Les Feuilles d'automne* in 1830, but less pronounced and below the level of significance). Three years after his first period ends, his epic novel *Les Miserables* (1862) would appear in print.

There is, to be sure, plenty to debate with this model. It is merely a first step in beginning to make our judgments explicit when thinking about the segmentation of the work of poets according to different moments within the author's life. Foote novelty is one possible way to do this. A great deal of work also remains to include more data and to ensure better representations of a poet's "collected works" in the data we already have.

But the point here is that with computational models we *can* have these debates—the terms and the points of reference are all out in the open on the table, cracks and all. There is a collective impulse at work here that is different from the ethos of traditional criticism, with its emphasis on the novel insight by an individual reader. More importantly, these models allow us to begin to compare poetic habits across a much broader range of authors and periods, to study something like Bourdieu's notion of a *habitus*, that set of shared stylistic practices within an author's life, and to identify when they give way to either temporary openings, as in the measures of local vulnerability, or more sustained windows of poetic change, as in the case of the period estimations or the fissures of global vulnerability. They allow us to see both the personal and historical factors that inform poetic change—as when we see Victor Hugo moving into a new poetic genre after his political exile or Émile Verhaeren reacting to the First World War—as well as some of the biology and corporality that surrounds creativity. These measures can give us a new sense of "corpus poetics"—the complicated relationship between embodiment and form.

Late Style

In his posthumously published book *On Late Style: Music and Literature Against the Grain* (2007), Edward Said crystallized long-standing, yet never entirely coherent, beliefs in the field about the uniqueness of artistic practice late in life.[20] For Said, whose book was perhaps not coincidentally deeply informed by Adorno's own late writings on late style—lateness twice over—late style was not so much a biological condition as it was an intellectual disposition of asynchronicity with one's own time. "Lateness" was not exclusively a period in the individual's life, the creative equivalent of senescence, but more a state of mind predicated on

CHAPTER SIX

anachrony, "the idea of surviving beyond what is acceptable and normal." Lateness was about "going against," a form of exile and estrangement that was nevertheless conditional upon biological aging.

Said's work gave voice to widely held assumptions by critics that as artists near the end of their lives their work changes in distinct ways. Jean Miró turns to bronze, Thomas Mann turns to pastiche, Beethoven becomes extremely challenging, and of course there is always *Faust 2*. For at least a sizable portion of artists, we imagine this turn to be one toward greater experimentation and difficulty. Freed from the exigencies of a career, and achieving a high point of synthetic power, much of the reception of late work focuses on this notion of stylistic resistance (but also grandiosity). While critics have reacted strongly against the singular nature of Said's formulation, his interest was not principally in defining late style more generally, but in understanding a particular kind of late style that he found in certain artists.[21] Nevertheless, the question remains as to whether there is something distinct or common across the modes of expression that artists choose as they age. While it is not possible to definitively answer that question here, I want to begin to formulate some approaches that might be able to give us insights into just how common the qualities of lateness are, at least for poets.

For the purposes of what follows, I will be defining "late period" as the final quarter of a poet's output. In this sense, it serves here less as a biological model and more as a creative one. I am interested in seeing whether the latter portion of a poet's writing has distinctive qualities, as opposed to testing for whether poets change based on their specific age (though this too is of course interesting). This approach has the advantage of allowing us to compare standardized portions of writing across different poetic careers. Given that poets vary considerably in their output over time as well as in the length of time that they live and work, not to mention when "old age" might be said to begin in any consistent way across time and different individuals' personal physiology, looking at lateness in this way allows us to observe it as a constant proportionality of a poet's creative work. We need a category that is sizable enough to observe a significant amount of output but that is also distinctively smaller in scope than the rest of the career. The "quartile" has served as a long-standing framework in statistical modeling, and thus seems to offer a useful starting point here.

The suite of measures I would like to examine are also bound together by their alignment with Said's notion of challengingness. They are designed to capture this sense of lateness, not simply as one of difference, but as a turn toward what I define as *the difficult, the irregular, the diverse,*

Table 6.2 Measures of late style (percentages represent the number of poets in the collection who exhibit that feature to a significant degree in their late work when compared with the rest of the career)

Feature	% poets	% that increase	% that decrease
Difficulty	53%	30%	23%
Syntactic irregularity	47%	37%	10%
Vocabulary richness	45%	30%	15%
Concreteness	33%	10%	23%
Generality	15%	6%	9%

the abstract, and the general. Taken together, they are meant to test some of our elementary assumptions about how poets create in the latter period of their lives. In table 6.2, I list the five measures used here as proxies for these ideas and the extent to which they rise or fall in a significant fashion in a poet's late work.[22]

We see for example that just over half of all poets in our collection (about 53%) show a significant increase or decrease in the "difficulty" of their writing in their late period, as measured by a standard readability score such as Tuldava's Text Difficulty Formula, which looks at sentence and word lengths.[23] For about half of all poets, there is some marked shift in the degree of difficulty, whether toward becoming more or less complex. Within this group, a majority show an increase in difficulty, suggesting that when there is a pronounced shift in language use late in life, the tendency is more often than not to become more difficult. Poets like Henri Michaux and Nelly Sachs become increasingly more difficult in their late work, while poets like Sarah Kirsch and Martine Audet become noticeably less so as they age.

Difficulty measured in this way captures the nature of vocabulary being used and the length of the sentences in which those words are appearing. But we can also look at a quality like syntactic irregularity, where we observe the relative rarity of certain syntactic patterns in poems and whether they appear more often late in life.[24] Here too we see just under half of all poets showing a significant change in their syntactic patterns later in their lives, with a majority once again showing a trend toward increased syntactic irregularity. It appears that when there is a shift, the tendency is not to move into a more serene, reduced, simplified mode of expression, but to become more stylistically challenging. Indeed, the window of the 10 most syntactically irregular poems in a poet's corpus is 54% more likely to occur in the late period than in the next most common quartile (the first). The most irregular late periods belong to the

CHAPTER SIX

contemporary Montreal poet Martine Audet and the modernists Renée Vivien, Louis Aragon, and Stefan George.

Another way to observe poetic change is to test whether there is a shift in vocabulary richness across the early and late periods, that is, whether an author gravitates toward a more diverse vocabulary as she or he ages.[25] Ian Lancashire famously used this measure to test whether a writer like Agatha Christie, who was diagnosed with dementia later in her life, saw her vocabulary richness decline in significant fashion.[26] The question here is whether late style implies a distinct shift to a more pared-down or more expansive vocabulary. Once more, we see just under half of poets exhibiting a change later in their lives, with a majority of those expanding their diversity of language late in life. Poets like Paul Celan, Rina Lasnier, and Else Lasker-Schüler significantly expand the diversity of their language as they age, while poets like Sylvia Plath, Muriel Rukeyser, and Alfred Lord Tennyson noticeably contract it. Plath and Celan are unique for each of their groups, with almost twice the change in vocabulary richness than the next highest person has in their respective categories—Celan becomes dramatically more expansive, while Plath shrinks her vocabulary in pronounced ways. The extremity of these poetic shifts offers a semantic insight into two poets who would end up taking their own lives.

When vocabularies change, how can we gain an understanding of the kinds of words that are changing? Here I use two separate measures to better understand the nature of the language being deployed in poets' later periods. The first explores whether the language becomes more "concrete," here defined as an increased prevalence of physical objects rather than cognitive abstractions, as defined by the WordNet taxonomies that I discussed in chapter one. A word like *vale* or *shore* will be labeled as an object, while *love* or *thought* as an abstraction. Their ratio, as well as its change in the late period, will tell us something about a poet's orientation toward concreteness or abstraction.[27]

The second measure looks at "generality," which considers the extent to which words in the late period are more likely to be hypernyms of words in the early period or vice versa. A hypernym we remember from chapter one is a more abstract representation of a similar entity ("furniture" is a hypernym of "chair"). We are asking whether the entities that populate a poet's late work tend to be more general than the rest of the career.[28]

Looking at the first measure, we see how fewer poets seem to show a significant change in the ratio of concreteness in their work (only about 33% indicate some kind of significant change). Of these, twice as many tend to *decrease* their reliance on physical entities as they age, or said another way, their language becomes significantly more reliant on abstractions. Poets like T. S. Eliot, Nelly Sachs, and J. W. Goethe fall into this group, while poets like Annette von Droste-Hülshoff, Joseph Eichendorff, and Mary Robinson belong to a rarer group who become significantly more oriented toward physical objects in their late work. Finally, we see how "generality" shows little significance across the corpus, with just twelve poets indicating a significant change in the degree of hypernymy in their late work (including Jean Cocteau, Sylvia Plath, and Christina Rossetti, who become significantly more "general").

As with the vulnerability scores discussed above, these measures are a beginning point to think about what it means to become more challenging in one's late period. There are certainly many more we might devise in the future, just as the ones on display here could undoubtedly benefit from more reflection and testing across a number of different scenarios. But when we look at them together, we see a relatively clear picture emerge about what we can and cannot say about Said's Adornoian theory of late style. The most important point is the way we do not see a strong orientation toward a unified late style. Only between a third and a half of all poets exhibit significant change in their late period according to any single measure. And only about a third show significant change for at least two measures. Lateness, according to this view, appears to be far from universal.

However, when we do see significant change in a poet's career, it is the case that for every measure but one the majority of poets will move toward greater challengingness: more difficulty, more irregularity, more vocabulary diversity, and more abstraction (though not more generality). While not all poets who embody this sense of lateness will engage in all of these practices simultaneously, when they do engage in any one of them it is far more likely that he or she will choose the more difficult or abstract path than the more straightforward, pared-down one.

Finally, this data can also give us insights into the different kinds of late styles on offer. As Linda and Michael Hutcheon have argued, rather than think in terms of a singular sense of challengingness, lateness can take many forms.[29] Not all of the poets engage in lateness in the same way, just as many do not show any significant change at all (William Carlos Williams, Friedrich Schiller, Muriel Rukeyser, or Marie Noël, to name a few). The graph below gives us an idea of the different directions that late style can go (fig. 6.8). It represents a biplot, which shows the relative positions of different poets according to the five dimensions of late style used here. Only poets who exhibit significant change in their late period according to at least two features are shown.

CHAPTER SIX

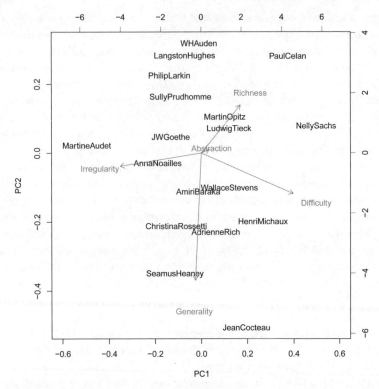

6.8 Biplot of the relationship of poets' late periods to features of stylistic difficulty.

In layperson's terms, the arrows indicate which variables are doing more of the work of accounting for the poet's location in the graph.[30] Where a poet is located indicates which variable or variables the poet manifests in a stronger way. The late work of Paul Celan, for example, is noticeable for its increased vocabulary range, while Nelly Sachs combines an investment in both an expanded vocabulary and greater stylistic difficulty. Seamus Heaney and Jean Cocteau become significantly more general, while Auden and Langston Hughes exhibit a combination of vocabulary richness plus syntactic irregularity. This is yet another way we can represent the distinctiveness of poets' careers as careers.

The Vulnerabilities of Wanda Coleman

Much of Said's thinking about lateness was informed by Adorno's modernist investment in the challengingness of the work of art. "But what of artistic lateness," Said asks at the outset of his book, "not as harmony and resolution but as intransigence, difficulty, and unresolved contradiction?" There is a kind of heroism or stoicism to late style, a Bartleby-like "I prefer not." "Lateness," writes Said, "is a kind of self-imposed exile from what is generally acceptable, coming after it, and surviving beyond it." And yet, given the far from universal nature of this model, we might want to ask what other ways we can think about lateness. How might we reframe Said's interest in lateness not as a form of heroic resistance but, following my earlier question, as one of poetic vulnerability, as an opening up to the not-yet-said of one's own life? As we have seen, aligning lateness with difficulty misses a great deal of the vulnerability, and the difference, at work in poets' later work.

A poet whose work offers an excellent example of this can be found in the career of Wanda Coleman (1946–2013). Coleman was a well-known African American poet, who felt that her work never achieved the same iconic status as other black writers of her era. As of today, there are fewer than 28 articles, dissertations, or book chapters about her work, compared with 168 for Maya Angelou and over 384 for Amiri Baraka (not to mention 1,410 for a white poet like W. H. Auden).[31] Her work can be harsh and hard hitting, but it is some of the most verbally dexterous and evocative writing I have ever read (she has the fifth most diverse poetic vocabulary of all poets in the collection).[32]

According to the measures of late style above, Coleman's poetry is not marked by a concerted investment in challengingness. While her vocabulary will become more diverse as she ages, the other measures either do not change or become less strong as she ages. What does change is the overall stylistic content of her poems. As you can see in figure 6.9, more than three-quarters of her vulnerable poems occur in the final quarter of her career; indeed, almost all of them occur after the publication of *Bathwater Wine* in 1998. As Coleman writes about herself, "By the end of 1996 everything was in shambles and 32-years of sacrifice to become a writer was tantamount to nothing."[33] Three years earlier she had published a collection of poems, *Hand Dance*, and felt stung by its lackluster reception.

We can see how *Bathwater Wine* marks out strongly new expressive terrain for Coleman, a move that is partially continued into her final volume of published poetry shortly before her death. She reacts powerfully to this sense of late malaise, or even failure. But she does so not by becoming more difficult or hermetic, by closing herself off from the world. Instead, she moves in new and unexpected ways relative to herself.[34]

What makes Coleman's work different after this point in her life? Once more, computation can be of help to ground our observations

CHAPTER SIX

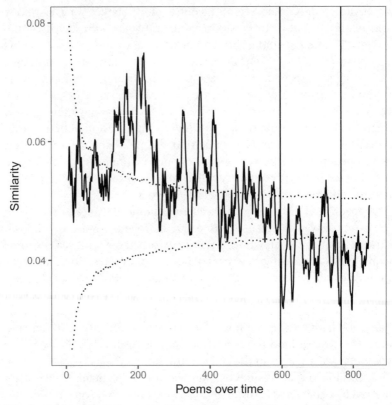

6.9   Local vulnerability in the poetry of Wanda Coleman. Vertical bars represent the poems contained in the collection *Bathwater Wine* (1998).

about these differences. The first thing we notice is that her sentences will become significantly shorter (though still not short by comparison), with the average per-poem sentence length dropping by 8 words in the vulnerable poems.[35] At the same time, there is a significant increase in punctuation in these poems (by about 70%).[36] But she initiates a vast new vocabulary as well. Over 35% of the words in her late vulnerable poems are not in earlier poems (more than 2,400 new words in 188 poems, or about 12–13 new words *per poem*).[37] If you were to take 1,000 random samples of a similar-sized group of poems from across her corpus, you would *never* get this much vocabulary shift. In other words, Coleman's vulnerability is marked by opening herself up to an enormous array of new words.

Capturing the specific nature of these words is challenging. Like all of her poetry, they are from high and low registers, *effluvium* and *blab-*

fest, incontrovertible and *piss-ant, psychosocial* and *pseudobrutha*, almost always multisyllabic and vibrant, like *sweetmamas, chaingangs, crackerjack,* and *cryptogram*. There are significantly more six-letter words, suggesting a syllabic elongation of her language. There seems to be a good deal more blood-talk, too, as in *bloodwart, bloodfree, blood-splashed, bloodstream,* and *bloodjoy*, as well as a preference for negative forms, like *impenetrable, implacable, improper, incontrovertible, incorrigible, indifferent, unconscious, unconvincing, undeserved, undone, unearthed, uneven, unholy, unnecessary, unsure, unstable,* and *unsettling* (there are more).

One of the biggest differences between the vulnerable and earlier poems is her reliance on nominalization. Overall we see a relatively strong increase in the use of plural nouns and proper names (1.36 times and 2.3 times more, respectively).[38] This nominal turn is not an entry into greater abstraction, however, a preference for the philosophical. Rather, it is marked by a significant decrease in abstraction, as her language becomes more physically concrete (though no less general).[39] Where her earlier, non-vulnerable poems tended to be more abstract, in her vulnerable poems this ratio flips so that there is an increased prevalence of physical things. There is a kind of deictic affirmation—a pointingness—to her late poetry that seems an essential part of her response to a sense of feeling adrift. This comes through most strongly in the increased presence of the "existential there" ("there is," or sometimes "there be," in Coleman) that appears significantly more often in her late work.[40] It is as though the naming and the negations combine to capture a feeling of what it means to open up, to enter into this new poetic space. The one assures, tries to get to the bottom of things, while the other takes away, doubts, says this is not possible. As Coleman writes in her sonnet "Morning Side Park" (1998) about visiting her father's gravesite:

there is no marker
but a small white
card protected in
see-thru plastic trimmed in green. his name
is typed on it and his date
of death. i am angry because i can't
afford to buy him a stone to equal
the weight of my love

Like all funereal scenes, the overall experience here is one of absence. But instead of focusing on the deceased parent, she focuses instead on the absence of an appropriate marker, the failure of consecration. The "i can't" stands out at the end of line twelve, which in the sonnet marks the moment just prior to what would be the Shakespearean couplet, the moment of resolution. "There is" only affirms something negative. It does not convey a sense of agency (in the sense of "prefer not," a sign of Said's late challengingness). Instead, it communicates a failure of agency, a poignant sign of vulnerability. This is what Levinas suggested in his ruminations on the impersonality of the "il y a," the way our recourse to affirming something's presence (something is there, there is) betrays a sense of some final absence.[41] Looking at the plastic-covered, typewritten note in lieu of a gravestone becomes a sign of her writing's failure more generally, the inability to find the right signs: "i am angry because I can't."

One can contrast this sense of failure, what she elsewhere calls "overwrought self-reproaches" and "doom's language my fatal folly," with the surreptitious pleasure of the hidden thing. In "Bad Eyes & the Wrong Prescription," Coleman paints a portrait of a woman peering behind "the forbidden door," behind which hangs "nighties of mauve, carnation and aquamarine / B-cup." It is a space of memory (bound letters, old cans, and handguns), but also transgression:

> and there,
> on the top shelf, an auburn wig for
> San Francisco sojourns at Beatitude and Fell.

It is a short poem about Tenderloin excursions, forbidden zones, and the warmth of recalling their passage, but most of all the things ("molting rabbit fur," "foot-smudged suede pumps a half toe too large," "silver-sequined stiletto-heeled slip-ons") that all run together in lavish, alliterative, dash-filled description.

In a brilliant essay on the fleshly nature of Coleman's writing, Rebecca Couch Steffy contrasts the idea of the "flesh" with that of the "body," seeing in flesh a more primordial material state, one that precedes all the ways bodies are caught up in social and legal norms.[42] Flesh is to some extent before the law, an act of grounding. But in these later poems of things, we can see all the ways that such grounding isn't about producing a moment of stability or recuperation, what Steffy calls "a fundamental human desire for recognition and fulfillment." Instead, it is about a relationship between concealment and communication. The poem ends with an acknowledgment of "betrayal," the way these things, and looking at them, betrays something that was otherwise hidden from view.

her forbidden door is made of clocks
whose furious kookoos reveal those lust-wrecked orbs
whose greedy chortles covet each ruby savory
betrayed by secret winks

This crazy door of "lust-wrecked orbs" possesses too much information (and too much possession, whose, whose . . .), chortling and winking at the same time. Coleman's poetry is about these things that are less familiar, but no less important, to the canon of English poetry, the way finding these things betrays some kind of knowledge, lets it out into the open. It crosses a line or a contract, but in doing so it communicates something vulnerable and hidden. The wink, the eye that while closed conveys that something untoward has been seen, encapsulates Coleman's late poetic vulnerability, its oscillation between, or perhaps better, its investment in, naming and undoing. It doesn't get to the bottom of things, but hovers among these strange, overripe configurations of objects.

Coleman's work is just one example of the way poetic vulnerability can function as a form of late style. As we saw above, there is no one strong tendency or set of tendencies that seems to encapsulate the numerous ways that creativity and aging might intersect. Where Said was drawn to the challengingness of late work, the way it walls itself off from the world as a form of resistance, I am far more drawn to a sense of vulnerability, seeing how poets open themselves up to a sense of something new and unknown. For Coleman, it means discovering not just a whole new range of things, but also the power of things to both convey and conceal. Vulnerability is about acknowledging the knowledge that is buried behind deranged kookoo clocks who wink back at us.

Conclusion (Implications)

"Up till now a few things still had to be set by hand, but from this moment it works all by itself." FRANZ KAFKA, *IN THE PENAL COLONY*

To be implicated in something is rarely seen as a good thing. "If they wanted to avoid pain and injury," writes one recent historian, "they would have to implicate not only themselves but others." "Joseph Monfre," writes another, "had been implicated in everything from the Lamana kidnapping of 1907 . . . to the Black Hand grocery bombings of the same year." Or: "Evidence implicated Berner, who confessed six times to six different people."[1] Building a vector space model of contemporary history writing confirms that the semantic associations of being implicated today are overwhelmingly negative.[2] The most similar words to *implicated* are *alleged, murder, suspect,* and *complicit.*

This was not always the case. One can find examples in different times and genres, where implication was more a fact of form: one spoke of the "implicated thorn" or "Grecian helms and implicated shields" or Nature's "implicated size," as in

(Where Nature wantons in minutest room,
Where folded close, her implicated size
Of trunk, branch, leaf, and future semen lies) [. . .][3]

To be implicated was aligned with its more literal meaning of being folded into something (*im* + *plicare*). It was an expression of interwovenness, but also something to come. "The boughes and armes of trees twisted one within another, [and] so implicated the woods together."⁴ It is telling that over time this sense of future entanglement would gradually come to be seen as one of the ultimate perils of our society. What could be worse than being subsumed within a larger plot, ceding our autonomy to the social implications around us? And yet what could be more beautiful than those implicated oaks, shields, boughs, and even thorns?

Throughout this book I have tried to show the ways that we can become more implicated in our observations about the literature of both past and present. In the act of modeling, we cede ourselves, fold ourselves into, the techniques and technologies through which we arrive at knowledge of our objects of study. We account more explicitly for the mediations that govern our insights. This is one way we become implicated in our scholarship.

But being implicated in something also entails acknowledging our agency in the process. It entails an act of recognition or accountability. As Marjorie Levinson wrote some time ago in response to the rise of New Historicism, "It is precisely our failure to articulate a critical field that sights us even as we compose it, that brings back the positivism, subjectivism, and relativism of the rejected [old] historicist methodology." In distancing ourselves from this past, says Levinson, "we construct for ourselves an experience of freedom and power with respect to our negotiations with the past."⁵ For Levinson, it is precisely our inability to account for ourselves in the analytical process that empties our work of its critical force.

By way of conclusion, I want to gesture toward the different ways we can begin to implicate ourselves more fully in our scholarship. It is a first attempt at undoing the ideology of the "by itself [*ganz allein*]" that was articulated in Kafka's fable of technology cited in my epigraph and that had undeniably pernicious outcomes. Whether it is the imagined autonomy of algorithms or that of the critical mind in exile, Kafka's text highlights for us the consequences of failing to acknowledge our own implications.

Self-Implication

One of the very first things that data allows us to do is assess ourselves. How can we begin to subject our own books and articles to the same scrutiny we level at those around us? How can we reduce that experience of freedom and power from the past that Levinson cautioned against? Here I introduce a few self-implication measures that focus principally

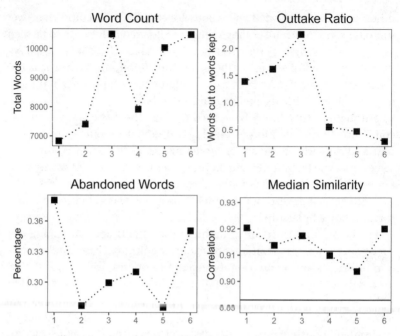

7.1 Four self-assessment measures that address the manuscript's length, revisions, and similarity to the field.

in two directions: observing our own revisionary practices and observing our similarities to the field.[6] Figure 7.1 shows four possible measures applied to each chapter of my book: word count, chapter-to-outtake ratio, percentage of abandoned words, and the median similarity of each chapter to a sample of JSTOR articles published in the past decade in the field of literary studies (the horizontal bands represent the confidence interval for a random sample of articles, what we would expect the average similarity to be between any given article and the larger field).[7]

We can see how there is some significant fluctuation in the chapter lengths, and that as the book progresses the amount of outtakes relative to the chapter's length decreases considerably. I generally keep everything I cut from a chapter while I edit, and so this measure gives me an idea of how difficult a chapter is to write—more outtakes are usually an indication that it took longer for me to figure out what I wanted to say. We can see how this process peaks in chapter three and drops off considerably, so that by the latter three chapters there are always fewer words cut than appear in the chapter. The progression mirrors in many ways the book's composition, where the further I moved along in the project

CONCLUSION (IMPLICATIONS)

the more I understood what I was doing and why. It would not surprise me if your own reading experience mirrors this trajectory, with a sense of the increasing quality and coherence of the chapters as the book progresses.

Abandoned words is another way of understanding what is happening during the process of revision, in this case how many words appear in the outtakes that do not appear in the chapter at all. I think of these as rejected alleyways of thought. While chapter six has the lowest amount of revision, it also has the second highest amount of abandoned words, suggesting that there were undercurrents in this chapter that disappeared (rather than get worked over and over). Finally, we see how most of my chapters tend to be a little more similar to JSTOR articles than the average JSTOR article. This surprises me, as I would have assumed that given the computational focus of the book there would be considerably more dissimilarity. Some of this may be due to the fuzziness of the model, but it might also suggest how squarely situated this book is within the mainstream of literary studies. It confirms that my primary effort here has been bridge-building rather than moving into entirely new disciplinary terrain.

Another way we might understand our implication in a field is through the semantic distinctiveness of our work. What words stand out in my chapters when compared to a language model built on articles from the field?[8] Figure 7.2 gives some indication of these words. In general, my writing in this book tends to focus on more general categories, like topic, novels, words, features, and characters, but also more plural

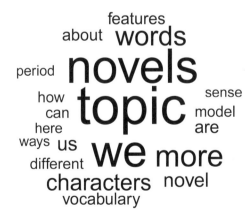

7.2 Most distinctive words of this book, compared to a sample of 2,000+ articles in literary studies from JSTOR.

181

categories (not to mention the notion of *more*). There is an investment in scale that makes this book unique, but also an investment in community: *we* and *us* are significantly overused here, giving you some insight into how I think, for better or worse. The two most under-used words in comparison to the JSTOR corpus are, interestingly, two more pronouns: *he* and *his*. This book is distinctive compared to the field because it does not over-invest in male authors or characters. These models can be a useful way of assessing whether we are reproducing biases intrinsic to the field or whether our work, at least semantically, is pushing research in new directions.

Finally, one method I use but will not pursue here is to cluster my chapters along with the articles from the field to discover new connections with existing research.[9] Using a vector space model, we can identify those articles or book chapters that are the most semantically similar to our own work beyond single keywords, titles, or author-as-subject fields. We have relied for far too long on search queries that are one-dimensional in nature. Further research is needed to better understand how searches based on multidimensional models perform and what scholarly value they add. Doing so needs to be part of our agenda to better understand how our writing is "sited" within the field, to use Levinson's well-chosen word.

Discipline and Distinction

These are just some of the ways we can begin to implicate ourselves as individuals within our research. But we can also undertake the same process on our field as a whole, to understand how we are implicated more collectively in the production of knowledge. What does all of our research, together, amount to?

This work traditionally has gone by the name of the "science of science," where computational models are used to understand the vast research outputs of scientific disciplines.[10] One of the reasons the humanities have been left out of this process is infrastructural—we have not developed the same kind of archival and indexing practices as the sciences, and thus do not have the same resources available for self-assessment. This is its own failure of implication and will require many years to correct.

Nevertheless, there are examples of this work already underway. Using the JSTOR for Research platform, Andrew Goldstone and Ted Underwood have developed a topic model of 21,000 articles from seven leading journals in the field of literary studies that stretch back over a century.[11] Mark Algee-Hewitt is developing methods of sequence alignment to study citational practices of nineteenth-century literary reviews in periodicals from ProQuest's British Periodicals collection.[12] These approaches can give us some idea of the semantic and citational attention of the field over the past two centuries, from the institutionalization of literary criticism in the nineteenth-century periodical press to its industrialization in the postwar academy.

But we can study the distributions of attention in our field using bibliographic indices as well, such as the MLA Bibliography. In 2015, for example, 20% of authors listed as subjects in the MLA bibliography accounted for just under 60% of all articles or book chapters published that year.[13] Just the top 1% of authors, or 33 in total, accounted for 1,302 essays, or one-fifth of the total. Four of these authors were women, and one was not white (W. E. B. Dubois). Those numbers are slightly more concentrated than in 1970, when 1% of authors accounted for 15.9% of all articles and book chapters. In that year, only one of the top-mentioned authors was a woman (George Eliot), and all were white.

Despite the consistency of such concentration, we are indeed talking about a considerably expanded range of authors today, up from 709 in 1970 to over 3,000 in 2015. The number of authors who are the focus of a scholarly article has measurably expanded. This means that each author now accounts for a smaller share of the overall total number of articles. Shakespeare has declined from being the subject of 7.5% of all articles in 1970 to about 3.4% today. Those who fear a dilution of the literary canon are thus partially correct. And yet the disproportionate nature of attention has changed little. The focus on the top looks much like it did close to a half century ago. Shakespeare was 4.1 times more likely to be the subject of an article than the next most discussed author in 1970 (John Milton), whereas he is now 4.5 times more likely to be discussed today (relative now to Charles Dickens). In other words, those who fear that the canon continues to dominate literary studies are also partially correct. If authors were like income, they would have a Gini coefficient of 0.51, down slightly from 0.53 in 1970 (for comparison, the United States has a coefficient of 0.45 for family income).[14] Like many other forms of cultural attention, authors continue to follow a very unequal distribution.

But we can also study our own authorial positions within the scholarship, once again siting ourselves within the larger field of academic publishing. As Chad Wellmon and I have recently shown, there is a tremendous bias in literary studies toward a handful of elite institutions that are responsible for producing much of the scholarship in leading

CONCLUSION (IMPLICATIONS)

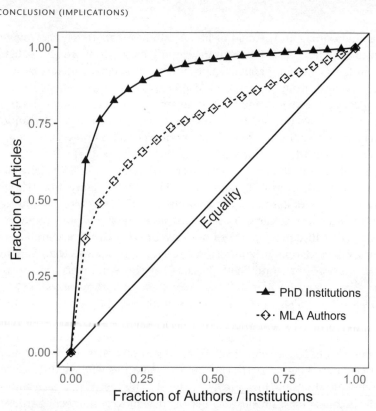

7.3 Lorenz curves comparing the relative inequalities surrounding how many academic articles are accounted for by the fraction of either PhD institutions or authors listed as subjects in the MLA Bibliography.

journals of the field.[15] When we examined the institutional affiliation of where authors were trained who had published in over 5,000 articles in four major literary studies journals (*PMLA, NLH, Representations,* and *Critical Inquiry*), we found that the top 20% of PhD-granting institutions accounted for 86% of all articles published since 1970.[16] The top ten institutions, which represent less than 3% of all institutions in our data set, accounted for just over half (50.7%) of all articles published, while just the top two, Harvard and Yale, accounted for one-fifth of all articles. The overall distribution of PhD institutions yields a Gini coefficient of 0.82. While these numbers decrease when we look at the author's institution of employment at the time of publication, there remains a considerable amount of inequality (Gini = 0.75).[17] Indeed, the inequalities surrounding our professional affiliations strongly outweigh those of our focus on the primary authors we choose to study (fig. 7.3). The *institutional* canon, in other words, is the stronger and more durable infrastructure of our field than the literary canon.

One of the questions we have been asking ourselves is what a notion of epistemic equality might look like. What would it mean to have a more inclusive academic system, one that is more responsive to a broader range of voices, including those outside the academy? And how might we design for it? It is clear that all of our recent calls for inclusivity have done little to undo a system that is highly unequal when seen from an institutional perspective. The current methods of selection surrounding hiring, promotion, and publication have reenforced, rather than overturned, long-standing institutional imbalances.

The argument this book makes is that data can help us see these imbalances, to make these latent institutional structures more visible. But it can also help us do something about them. Through this work we are trying to imagine new forms of algorithmic openness, where computation is used not as an afterthought—as a means of searching for things that have already been preselected and sorted—but as a form of forethought, as a means of generating more diverse ecosystems of knowledge. What values do we care about in terms of human knowledge, and how can we use the tools of data science to capture and more adequately represent those values in our systems of scholarly communication? Instead of subject indexes and citation rankings, one could imagine filtering by institutional diversity, citational novelty, matters of public concern, or any number of other priorities. How might we *encode* these values to create smarter, more adaptable, and more inclusive publishing platforms?

We now have a variety of new tools and new resources at our disposal that can help us better understand our implicatedness in our research. The only thing standing between us and knowing more is our imaginations.

Acknowledgments

If this book did not exactly start out in a garage, it can reasonably be said to have begun in a studio in Montreal's Old Port. This is where my colleague Mark Algee-Hewitt, then a postdoc at McGill University, sat me down one day at his terminal and started teaching me how to program in order to study literature. I no longer remember exactly what we were working on (*Werther*, maybe). But it was that spirit of taking the time to make this work possible for someone else, to help people see what you see, that I have taken as my guiding model ever since. This book owes a large debt of gratitude to Mark, not just because he helped me get started, but also because he has been such an inspiring interlocutor ever since.

Work in computational criticism (or cultural analytics) tends to be much more collective and collaborative than its bibliographic predecessors. A great number of people have played a significant role in helping bring this book to fruition. Richard Jean So and Chad Wellmon have been incredible collaborators and coauthors. They have helped me find my way to the intellectual pressure points of the discipline. I continue to be inspired by their leadership. I want to thank all of the members of the Novel™ research group, Anatoly Detwyler, Geoffrey Rockwell, Hoyt Long, Laura Mandell, Mark Algee-Hewitt, Matthew Erlin, Matthew Jockers, Richard Jean So, Stéfan Sinclair, Stephen Downie, Susan Brown, Ted Underwood, and Matthew Wilkens, who have formed a core community for me over the past few years where I have been able to try out new ideas and learn intensely from theirs. I also want to thank my two collaborators in computer science, Derek Ruths and Mohamed Cheriet, who have taught me a tremendous amount about how computer scientists think. And of course, thanks to my writing group, Jonathan Sachs and Omri Moses, who provide much-needed intellectual and not-so-intellectual camaraderie.

In addition to the work of colleagues, a great number of students have facilitated this work in numerous ways over the years. It's because of them that my lab is such a rewarding place to work. I can go there and forget about the squabbles of a gestating field and focus instead on building models, asking questions, and the personal rewards of mentoring new scholars. In particular, I would like to thank Fedor Karmanov, Eva Portelance, Eve Kraicer, Koustuv Sinha, Hardik Vala, Victoria Svaikovsky, Esther Vinarov, Stefan Dimitrov, Anne Meisner, Lisa Teichmann, and Shoshana Schwebel. Meng Zhao read through the entire manuscript to review the statistical models and came away a fan of cultural analytics. I am particularly grateful to the Social Sciences and Humanities Research Council for making the funding available to support so many student collaborators.

Once more, I would also like to thank my editor, Alan Thomas, who remains a continued source of sound advice and unwavering support. I also greatly appreciate the peer-review reports that helped strengthen the manuscript in numerous places as well as the book's more global message.

Acknowledgments are where we go as authors to pay our intellectual debts before we take all the credit. But to acknowledge something, as opposed to someone, is also to write in a confessional mode, to make some aspect of personal experience public. It would be disingenuous to pretend that this book simply "happened" like the others, and not acknowledge the very real challenges imposed by moving into new disciplinary terrain. I have experienced some of the most exciting moments of learning and discovery while writing this book, as well as some of the most disquieting moments of intellectual antagonism of my career (one example: "When I see those graphs, it makes me want to vomit"). It appears, at least for now, that you cannot have one without the other. Undoing long-standing conventions of disciplinary self-understanding also means losing colleagues, communities, and all of those invisible privileges that make academic labor flow more smoothly. For those thinking about entering the field, it is important to acknowledge this, too. I am by no means the first to do so. Many others have already gone through this and helped pave the way to get to this point. To them I am very thankful. I am sure many others will continue to move this kind of work forward. To them I am also extremely grateful.

———

A version of chapter four appeared in a slightly altered form in *Cultural Analytics* (December 2016), and the introduction represents a substantially rewritten account of the ideas initially developed in two pieces, "Reading's Refrain: From Bibliography to Topology," *ELH* 80.2 (2013): 373–399; and "There Will Be Numbers," *Cultural Analytics* (May 2016).

Appendix A

Table A.1 Distinctive features of twentieth-century poems with excessive amounts of periods. The top half of the columns represent those features that are over-represented, and the bottom half (bold) represent those that are under-represented.

	Words		Hypernyms	Parts of Speech (POS)
	a	me	communication	Personal Pronoun (PRP)
	I	this	instrumentality	Plural Noun (NNS)
	you	one	auditory	Proper Name (NNP)
	is	no	change	Verb Pres 3rd Pers Sing (VBZ)
positive	it	come	**attribute**	Verb Present Non-3P Sing (VBP)
	my	love	**measure**	Cardinal Number (CD)
	are	eye	**quantity**	Quotation (")
	no	am	**time_period**	**Noun Singular (NN)**
	your	know	**quality**	**Determiner (DT)**
	me	night	**water**	**Prepositions (IN)**
	the	**from**	**speech**	**Adjectives (JJ)**
	of	**was**		**Adverbs (RB)**
	and	**as**		**Verb Past Tense (VBD)**
	to	**but**		**Coordinating Conj (CC)**
negative	**with**	**or**		**Possessive Pron (PRP$)**
	that	**their**		**Verb Past Participle (VBN)**
	he	**by**		**to (TO)**
	his	**so**		**Modal (MD)**
	from	**where**		**Wh-Pronoun (WP)**
	was	**our**		**Possessive (POS)**

Table A.2 Most contractive novels—Model 1

English	z
Ann Radcliffe, *Mysteries of Udolpho* (1794)	−2.66
H. G. Wells, *The Time Machine* (1895)	−2.45
Mary Ward, *Robert Elsmere* (1888)	−1.89
James Joyce, *Ulysses* (1922)	−1.80
George Eliot, *The Mill on the Floss* (1860)	−1.72

French	z
Jules Verne, *De la terre à la lune* (1865)	−3.37
Jules Verne, *Vingt mille lieues sous les mers* (1870)	−2.72
Jules Renard, *Poil de Carotte* (1894)	−2.14
Nicolas-Edme Rétif, *Histoire de Sara* (1796)	−2.07
Louis Duranty, *Le malheur d'Henriette Gerard* (1860)	−2.01

German	z
Theodor Fontane, *Irrungen, Wirrungen* (1887)	−2.51
Friedrich Klinger, *Fausts Leben* (1791)	−2.48
K. P. Moritz, *Anton Reiser* (1785)	−2.20
Charlotte von Ahlefeld, *Marie Müller* (1814)	−1.98
Paul Scheerbart, *Lesabéndio* (1913)	−1.92

Table A.3 Most contractive novels—Model 2

English	z
Virginia Woolf, *Mrs. Dalloway* (1925)	−2.11
Ann Radcliffe, *The Mysteries of Udolpho* (1794)	−2.03
Jane Austen, *Sense and Sensibility* (1811)	−1.92
Mary Ward, *Robert Elsmere* (1888)	−1.83
Mary Shelley, *Frankenstein* (1818)	−1.63

French	z
Jules Verne, *De la terre à la lune* (1865)	−4.49
Jules Verne, *Vingt mille lieues sous les mers* (1870)	−2.19
Nicolas-Edme Rétif, *Histoire de Sara* (1796)	−2.01
Alphonse de Lamartine, *Le Tailleur de pierres* (1851)	−1.95
Etienne Senancour, *Obermann* (1840)	−1.64

German	z
Joachim Campe, *Robinson der Jüngere* (1779)	−2.12
J. W. Goethe, *Wilhelm Meisters Wanderjahre*	−1.84
Ilse Frappan, *Arbeit* (1903)	−1.76
Rainer Maria Rilke, *Malte Laurids Brigge*	−1.71
Adalbert Stifter, *Der Nachsommer*	−1.70

Table A.4 Novels that contain Topic 150 with more than a 20% probability. The column on the right presents the average probability of the segments along with the highest document probability in the case of there being multiple segments.

Author	Title	Date	Segments	Avg. (max) probability
Goethe, J. W.	Werther	1774	1	0.24
Moritz, K. P	Anton Reiser	1785	1	0.22
Moritz, K. P.	Andreas Hartknopf	1786	1	0.44
Naubert, Benedikte	Alf von Dülmen	1791	1	0.21
Jacobi, Friedrich	Eduard Allwill	1792	1	0.25
Paul, Jean	Hesperus	1795	2	0.25 (0.26)
Tieck, Ludwig	William Lovell	1796	4	0.36 (0.56)
Hölderlin	Hyperion	1799	5	0.25 (0.31)
Brentano, Clemens	Godwi	1802	1	0.25
Arnim, Achim von	Hollins Liebeleben	1802	2	0.22 (0.22)
Arnim, Achim von	Gräfin Dolores	1810	1	0.32
Fouqué, Caroline	Frau des Falkensteins	1810	1	0.20
Arnim, Bettine von	Goethe's Briefwechsel	1835	2	0.24 (0.27)
Gutzkow, Karl	Wally	1835	1	0.22
Achim, Bettine von	Clemens Brentano's Frühlingskranz	1844	2	0.39 (0.45)
Hahn-Hahn, Ida	Sibylle	1846	1	0.20
Frapan, Ilse	Arbeit	1903	1	0.22
Heyking, Elisabeth von	Briefe, die ihn nicht erreichten	1903	1	0.22
Rilke, Rainer Maria	Malte Laurids Brigge	1910	1	0.24

Table A.5 Data overview for the fictionality tests

Collection	Key	Description	Documents
	EN_FIC	English Fiction	100
	EN_NOV	English Novels	100
	EN_NOV_3P	Eng. Novels 3-Person	107
19C Canon	EN_NON	English Nonfiction	100
NOVEL450	EN_HIST	English Histories	85
NON_19C	DE_NOV	German Novels	100
	DE_NOV_3P	Ger. Novels 3-Person	110
	DE_NON	German Nonfiction	100
	DE_HIST	German Histories	75
	HATHI_FIC	Hathi Trust Fiction	9,426
Hathi Trust	HATHI_NON	Hathi Trust Nonfiction	11,732
	HATHI_TALES	Hathi Trust Fiction Minus Novels	428
1790-1990 NOVEL19C NOVEL20C	STAN_KLAB	English Novels	6,421
	CONT_NOV	Contemporary Novels	200
Contemporary	CONT_NOV_3P	Cont. Novels 3-Person	210
NOVEL_CONT	CONT_NON	Contemporary Nonfiction	200
NON_CONT	CONT_HIST	Contemporary Histories	200

Table A.6 The top twenty features with the greatest increase in fiction compared to nonfiction. Values represent median percentages of features for each collection.

	Fiction vs. nonfiction w/dialogue removed 19C canon (English)			
Feature	Category	Fiction (%)	Nonfiction (%)	Ratio
family	social	0.53	0.15	3.53
exclam	linguistic	0.07	0.02	3.50
shehe	linguistic	5.92	1.92	3.09
body	biological	1.28	0.50	2.56
home	social	0.71	0.29	2.43
sexual	social	0.18	0.08	2.19
see	perceptual	1.26	0.59	2.14
hear	perceptual	0.65	0.31	2.13
past	linguistic	6.49	3.11	2.09
feel	perceptual	0.77	0.37	2.07
friend	social	0.18	0.10	1.89
percept	perceptual	2.81	1.49	1.89
anx	affective	0.47	0.26	1.81
ppron	linguistic	8.73	4.90	1.78
bio	biological	2.23	1.27	1.76
ingest	biological	0.30	0.18	1.67
social	social	11.81	7.69	1.54
sad	affective	0.60	0.39	1.53
motion	relative	2.23	1.49	1.50
assent	oral	0.03	0.02	1.50

p-value < 0.0001

Table A.7 The top twenty features with the greatest increase in third-person novels compared to nineteenth-century histories. Values represent median percentages of features for each collection.

	Novel (3P) vs. history w/dialogue removed 19C canon (English)			
Feature	Category	3P novels (%)	History (%)	Ratio
exclam	linguistic	0.08	0.02	4.00
hear	perception	0.67	0.19	3.53
see	perception	1.30	0.37	3.51
feel	perception	0.84	0.26	3.21
percept	perception	2.95	0.92	3.20
shehe	linguistic	6.95	2.25	3.09
body	biological	1.27	0.41	3.09
sexual	biological	0.19	0.06	3.08
assent	oral	0.03	0.01	3.00
qmark	linguistic	0.09	0.03	3.00
home	social	0.73	0.26	2.84
anx	affective	0.56	0.24	2.38
bio	biological	2.25	0.98	2.30
ingest	biological	0.30	0.14	2.14
ppron	linguistic	8.42	4.17	2.02
sad	affective	0.69	0.36	1.92
friend	social	0.17	0.09	1.89
family	social	0.45	0.26	1.75
discrep	cognitive	1.27	0.74	1.72
social	social	12.04	7.00	1.72

p-value: < 0.0001

Table A.8 The top twenty features with the greatest increase in third-person novels compared to nineteenth-century fiction. Values represent median percentages of features for each collection.

Feature	Category	3P novels vs. other fiction 19C canon (English)			
		Novel (%)	Other fiction (%)	Ratio	*p*-value
qmark	linguistic	0.54	0.33	1.65	***
assent	oral	0.17	0.11	1.57	**
period	linguistic	4.92	3.82	1.29	***
present	linguistic	4.45	3.52	1.27	***
discrep	cognitive	1.61	1.31	1.23	***
shehe	linguistic	5.82	4.92	1.18	*
negate	cognitive	1.68	1.43	1.18	***
see	perception	1.16	1	1.16	*
verb	linguistic	12.91	11.17	1.16	***
percept	perception	3.14	2.74	1.15	**
hear	perception	1.12	0.98	1.14	*
insight	cognitive	2.15	1.9	1.13	**
tentat	cognitive	2.27	2.01	1.13	**
auxverb	linguistic	7.78	6.93	1.12	***
past	linguistic	6.2	5.54	1.12	**
feel	perception	0.68	0.61	1.11	.
future	linguistic	1.2	1.09	1.1	*
adverb	linguistic	3.74	3.41	1.1	*
excl	cognitive	2.26	2.09	1.08	*
social	social	13.29	12.37	1.07	*

p-value codes = *** < 0.0001, ** < 0.001, * < 0.01, . < 0.05

Table A.9 Distinctive vocabulary of nineteenth-century novels when compared to other fiction across four categories: *discrepancy, insight, tentativeness,* and *negation*. G^2 represents Dunning's log-likelihood ratio where a higher score represents an increased likelihood of a word appearing in fiction.

Discrepancy	Frequency (per 100K)	G^2	**Insight**	Frequency (per 100K)	G^2
want	84.11	2086.73	think	154.93	3241.93
would	348.14	1534.14	know	183.74	1564.75
if	315.85	1213.50	realiz	5.41	1524.74
wouldnt	13.15	978.85	thought	139.88	801.93
could	246.67	738.47	seem	120.01	730.36
couldnt	10.73	683.50	felt	64.81	557.08
shouldnt	5.09	638.09	sens	31.39	547.27
rather	51.37	472.47	feel	86.05	483.08
must	131.82	439.63	meant	15.03	465.98
mustnt	2.58	386.01	recogn	3.78	434.78
hope	68.91	362.06	conscious	18.16	422.69
ought	23.95	292.34	rememb	38.16	413.15
neednt	2.50	268.92	believ	57.45	369.91
should	148.47	175.96	idea	30.27	331.38
ideal	3.36	149.13	knew	65.63	318.91
wish	65.77	142.12	question	36.44	267.27
normal	0.85	136.66	understand	28.38	235.23
problem	1.99	112.16	decid	14.48	197.34
oughtnt	0.72	105.00	mean	64.27	162.91
need	28.78	99.08	percept	3.82	122.36
Negation	Frequency (per 100K)	G^2	**Tentative**	Frequency (per 100K)	G^2
dont	107.10	6643.46	like	195.81	1262.54
cant	34.32	2369.73	someth	62.90	1091.98
isnt	13.82	1375.74	quit	74.78	1003.91
didnt	21.55	1322.70	ani	162.81	829.51
doesnt	8.48	961.32	anyth	43.52	527.88
havent	7.74	834.73	vagu	8.38	470.24
wont	24.89	747.67	suppos	38.00	391.83
not	715.73	559.28	hope	68.91	362.06
never	149.50	414.82	almost	51.55	357.66
arent	2.41	378.80	hard	45.05	317.18
shant	3.19	376.61	perhap	48.76	311.51
no	332.10	341.80	sort	28.85	308.24
hadnt	5.22	324.28	might	118.27	269.19
wasnt	6.94	294.65	question	36.44	267.27
noth	86.00	268.27	guess	11.43	242.72
hasnt	2.51	233.40	possibl	36.26	180.39
werent	1.25	157.70	somehow	5.94	179.07
aint	6.25	58.15	chanc	19.99	142.57
nobody	10.45	49.32	anybodi	7.05	141.16
negat	1.25	44.62	anyon	3.27	116.36

Table A.10 Nineteenth-century novels by female authors with the most perceptually oriented female main characters

	Perception	
Novel	Date	z-score
Stoddard, *The Morgesons*	1862	2.32
Radcliffe, *Mysteries of Udolpho*	1794	2.14
Alcott, *Moods*	1864	2.01
Rowson, *Charlotte*	1791	1.90
Austen, *Pride and Prejudice*	1813	1.70

	Perception/cogitation ratio	
Novel	Date	z-score
Rowson, *Charlotte*	1794	4.90
Stoddard, *The Morgesons*	1862	4.89
Stoddard, *Two Men*	1865	1.61
Evans, *Inez*	1855	1.43
Holmes, *Elsie Venner*	1861	1.15

Appendix B: Model Description for Deriving Poetic Similarity

Definition Two poems are similar if they share a significant amount of features across four key dimensions of a poem that include the lexical, semantic, syntactic, and phonetic dimensions with priority given to semantic relations.

Lexical I define lexical features as word stems that have been normalized by tfidf score, meaning words are weighted by their relative rarity in a corpus (the less common a word is overall, the more important it is in defining whether two poems are similar). Contrary to authorship attribution approaches, I also remove stopwords and thus only consider words thought to have higher semantic significance.

Semantic I define semantic features as the transformation of lexical features using singular value decomposition as applied in Lexical Semantic Analysis (LSA). What this means in practice is the reduction of the number of words to more general semantic commonalities between them based on co-occurrences within poems. *Love* and *desire* are different words that capture similar semantic associations. The transformed representation should capture this semantic similarity where there is not an overlap at the level of the lexeme.

Syntactic I define syntactic features as part of speech (POS) ngrams, where I keep only 1–2 POS grams. This means each poem is transformed into its POS representation (noun, adjective, verb, determiner, etc.), with only single parts of speech and pairs retained. These are then normalized by tfidf score to find those usages that are comparatively rare in the corpus and weighted more significantly.

Phonetic   I define phonetic features as the transformation of words into their respective phonemes using the Open Mary speech-to-text software. Rhyme, assonance, and consonance are accounted for in a distributional sense by observing how often a particular sound recurs in a given poem. As with other features, phonemes are normalized by tfidf to capture those sounds that are less common across the corpus and weighted accordingly.

Similarity   I define similarity as cosine similarity between the distribution of features for any two poems.

Implementation   The model has two basic guiding principles. The first is the prerequisite of similar length. Poems can only be similar to other poems of similar length, where I define similar length as being no more than 100% longer or 50% shorter. (A poem of 70 words can only be similar to poems less than 140 words and longer than 35.) Second is the priority of semantic similarity. Rather than treat all features equally in a single representation, I prioritize semantic similarity by treating the semantic representation as its own feature space and use it to calculate the similarity between poems. I then combine this representation with the similarity scores between poems based on a *combined* feature space of lexical, syntactic, and phonetic features, which have also been adjusted for length (this cuts the correlation between similarity score and length in half). What this means in practice is that the semantic features have the strongest influence which undergo a slight correction with the presence of the other, more literal, features. As I discuss in the chapter, the reason for doing so is that the semantic associations identified by the model are quite powerful by themselves (especially when compared with the lexical), but will on occasion draw connections that are too associative. The lexical, syntactic, and phonetic features help reduce these misassociations by emphasizing the necessity of more literal, acoustic, and syntactic connections between poems. Finally, for the purposes of making choices about how to draw connections between poems, I take only the most similar poem unless there is an ambiguity among the top choices, in which case I take all poems until there is an unambiguous distinction between them. Ambiguity is defined as a difference between the nth most similar poem and the $n+1$ most similar poem that is *smaller* than the difference between the next two poems (i.e., $n+1$ and $n+2$). What I am looking for is a clear distinction between the similarity scores between poems. The result is usually in the range of 1–3 poems chosen for any given poem.

Data Sets

The following is a list of the data sets used in this book and their shorthand references. Metadata about the collections is available with the accompanying online supplemental material located here: github.com/piperandrew/enumerations.

Literature Online Twentieth-Century Poetry: 75,297 poems in English written by authors born between 1865 and 1975, drawn from the ProQuest Literature Online Collection. POETRY_20C.

Literature Online Nineteenth-Century Poetry: 125,675 poems in English written by authors born between 1765 and 1865, drawn from the ProQuest Literature Online Collection. POETRY_19C.

.txtLAB Collected Works of Poetry: 32,022 poems in French, German, and English representing 78 poets' collected works written between 1597 and 1946. POETRY_CW.

.txtLAB 450: 450 largely canonical novels written in French, German, and English published between 1770 and 1930. NOVEL450.

.txtLAB 19C: A collection of 361 works of nonfiction in English and German, including histories, memoirs, philosophy, and essays. NON_19C.

.txtLAB Contemporary Literature Collection: 1,209 novels in English published between 2005 and 2015 representing six different genres (Bestsellers, Prizewinners, Mysteries, Romances, SciFi, and novels reviewed in the *New York Times*) and 600 works of nonfiction. NOVEL_CONT and NON_CONT.

Stanford Nineteenth-Century Novels: 3,285 novels in English published between 1800 and 1899, collected by the Stanford Literary Lab. NOVEL_19C.

Chadwyck Nineteenth-Century Novels: A subset of 750 more canonical novels from the Stanford Collection. NOVEL750.

DATA SETS

Chicago Twentieth-Century Novels: 2,206 novels in English published between 1880 and 2000, randomly drawn from the full Chicago Text Lab collection. NOVEL_20C.

Chicago Modernist Novels: 1,000 novels in English published between 1880 and 1930, randomly drawn from the Chicago Text Lab collection. MODERN_20C.

Hathi Trust Nineteenth-Century Fiction: 9,426 unique volumes of fiction in English published between 1800 and 1900, drawn from the Ted Underwood data set. All duplicates and works with "tales," "stories," "essays," or "scenes" in their titles or genre description have been removed. HATHI_FIC.

Hathi Trust Nineteenth-Century Nonfiction: 11,732 works of nonfiction in English published between 1800 and 1900. HATHI_NON.

Hathi Trust Tales Collection: 428 works of non-novelistic fiction in English, including epics translated into prose, folktales, and domestic tales. HATHI_TALES.

Notes

INTRODUCTION

1. This number was calculated based on taking a random selection of 25,500 pairs of pages from 1,275 novels published since 1950 (NOVEL_20C). A page is defined as a standardized 500-word window of text. See script 0.1 in the introduction directory.
2. Rolf Engelsing's work on debates surrounding "extensive" versus "intensive" modes of reading ca. 1800 is foundational in this regard. See Rolf Engelsing, *Der Bürger als Leser: Lesergeschichte in Deutschland 1500–1800* (Stuttgart: Metzler, 1974). Such sentiments continue to be echoed in the slow reading or slow research movement today. See John Miedema, *Slow Reading* (Sacramento, CA: Litwin Books, 2009), and more recently Maggie Berg and Barbara Seeber, *The Slow Professor: Challenging the Culture of Speed in the Academy* (Toronto: Toronto University Press, 2016). Because we can never know when enough is enough—that is, when one has read too much (enough!) or too little (not enough!)—such a framework always requires some kind of apparatus of surveillance to maintain itself. On reading as a form of self-help, see Paul Reitter and Chad Wellmon, "Better Living Through Bibliotherapy," *Hedgehog Review* 18.2 (2016).
3. The foundational works in this area are Lucien Febvre and Henri-Jean Martin, *The Coming of the Book: The Impact of Printing, 1450–1800* (London: Verso, [1958] 2010); and D. F. McKenzie, *Bibliography and the Sociology of Texts* (Cambridge: Cambridge University Press, [1986] 1999). Other works that take quantity as the starting point of intellectual history are Martyn Lyons, *Le Triomphe du livre: Une histoire sociologique de la lecture dans la France du XIXe siècle* (Paris: Promodis, 1987);

James Smith Allen, *Popular French Romanticism: Authors, Readers, and Books in the Nineteenth Century* (Syracuse: Syracuse University Press, 1981); James Raven, *The Business of Books: Booksellers and the English Book Trade, 1450–1850* (New Haven, CT: Yale University Press, 2007); William St. Clair, *The Reading Nation in the Romantic Period* (Cambridge: Cambridge University Press, 2004); Ilsedore Rarisch, *Industrialisierung und Literatur* (Berlin: Colloquium, 1976); Ann Blair, *Too Much to Know: Managing Scholarly Information Before the Modern Age* (New Haven, CT: Yale University Press, 2011); and Chad Wellmon, *Organizing Enlightenment: Information Overload and the Invention of the Modern Research University* (Baltimore: Johns Hopkins University Press, 2015).

4. Gilles Deleuze, *Foucault*, trans. Sean Hand (Minneapolis: University of Minnesota Press, 1988).

5. Alan Liu, "Where Is Cultural Criticism in the Digital Humanities?," *Debates in the Digital Humanities*, ed. Matthew K. Gold (Minneapolis: University of Minnesota Press, 2012), 490–509. For work that emphasizes the interpretive aspects of data-driven literary study, see Geoffrey Rockwell and Stéfan Sinclair, *Hermeneutica: Computer-Assisted Interpretation in the Humanities* (Cambridge, MA: MIT Press, 2016).

6. A fuller description of the data can be found in the appendix. All data and code used in this book is available online through the link provided in the appendix.

7. I draw this term from the sociologist Arlie Russell Hochschild, who uses it in a different way in her book *Strangers in the Their Own Land: Anger and Mourning on the American Right* (New York: New Press, 2016). For Hochschild, the deep story is a narrative that precedes one's experience and interpretation of experience such that it shapes all incoming information within that narrative framework.

8. On the importance of pattern recognition and computational criticism, see Hoyt Long and Richard Jean So, "Literary Pattern Recognition: Modernism Between Close Reading and Machine Learning," *Critical Inquiry* 42 (Winter 2016): 235–266.

9. Given the many conflicting definitions surrounding what has come to be known as the digital humanities, I would advocate for the use of new terms to refer to the specific work of "computational criticism" or "cultural analytics." These can be seen as subsets of the larger initiatives within the digital humanities and refer to the practice of using computational methods for the study of literature or culture. See Andrew Piper, "There Will Be Numbers," *Cultural Analytics* (May 2016), doi:10.22148/16.006.

10. Several new projects are underway to reverse this trend to begin to write the history of data or information. For a project that focuses on the "big humanities" of the nineteenth century, see Chad Wellmon, "Loyal Workers and Distinguished Scholars: Big Humanities and the Ethics of Knowledge," *Modern Intellectual History* (2017): 1–37. Rockwell and Sinclair also con-

tinuously imbed computational literary criticism within longer historical contexts. See Rockwell and Sinclair, *Hermeneutica*.
11. See Chris Cannon, "The Art of Rereading," *ELH* 80.2 (2013): 401–426; and Deidre Shauna Lynch, "Going Steady: Canon's Clockwork," *Loving Literature: A Cultural History* (Chicago: University of Chicago Press, 2015), 147–194.
12. For new work that focuses on the computational study of reprinting, see Ryan Cordell, "Reprinting, Circulation, and the Network Author in Antebellum Newspapers," *American Literary History* 27.3 (2015): 417–445.
13. There is a strong affinity in this line of thinking with the post-structural emphasis on intertextuality that needs to be made explicit in new research using computational methods. Many of the basic priorities of a post-structural way of viewing texts are implicitly operationalized by computational models. Nevertheless, as I will try to show throughout this book, the ways in which computational models do so are far richer and more diverse than a notion of intertextuality understood, in Barthes's terms, as a "tissue of quotations" allows for. Seen computationally, the relationship between texts goes well beyond a purely citational model.
14. The research of Rita Felski is essential here for realigning our focus on the ways in which attachments and affinities are maintained over time, as is the work of Wai-Chee Dimock and her emphasis on "deep time." See Rita Felski, *The Limits of Critique* (Chicago: University of Chicago Press, 2015); and Wai-Chee Dimock, *Through Other Continents: American Literature Across Deep Time* (Princeton, NJ: Princeton University Press, 2006). For a discussion of the changing scale of literary analysis, see James F. English and Ted Underwood, "Shifting Scales: Between Literature and Social Science," *MLQ* 77.3 (2016): 277–295. For initial computational work on cultural stability, see Ted Underwood, "The Life Cycles of Genres," *Cultural Analytics* (May 2016), article doi:10.22148/16.005, dataverse doi:10.7910/DVN/XKQOQM; and Ted Underwood and Jordan Sellers, "The Longue Durée of Literary Prestige," *MLQ* 77.3 (2016): 321–344. For computational work on transcultural flows, see Hoyt Long and Richard Jean So, "Turbulent Flow: A Computational Model of World Literature," *MLQ* 77.3 (2016): 345–367. For work on forms of cultural distinction, see Andrew Piper and Eva Portelance, "How Cultural Capital Works: Prizewinning Novels, Bestsellers, and the Time of Reading," *Post45* (May 2016), http://post45.research.yale.edu/2016/05/how-cultural-capital-works-prizewinning-novels-bestsellers-and-the-time-of-reading/.
15. Friedrich Kittler, *Optical Media* (Cambridge: Polity, 2010), 230. For an attempt to think through the congruences between writing, images, and number in the German tradition, see Sybille Krämer and Horst Bredekamp, eds., *Bild, Schrift, Zahl* (Munich: Fink, 2003).
16. See English and Underwood, "Shifting Scales"; Richard Jean So and Hoyt Long, "Network Analysis and the Sociology of Modernism," *Boundary 2* 40.2 (2013): 147–182; and Ted Underwood, "A Genealogy of Distant Reading,"

Digital Humanities Quarterly 11.2 (2017), http://www.digitalhumanities.org/dhq/vol/11/2/000317/000317.html.
17. See contributions to the journal *Scientific Study of Literature* for research that focuses on a primarily social-psychological approach to the study of literature. Also under this umbrella is the new interest in neuro-aesthetics, for example in the work of G. Gabrielle Starr, *Feeling Beauty: The Neuroscience of Aesthetic Experience* (Cambridge, MA: MIT Press, 2013).
18. In this, this book can be seen as a more applied extension of existing introductory works to the field, such as Matthew L. Jockers, *Macroanalysis: Digital Methods and Literary History* (Chicago: University of Illinois Press, 2013); Geoffrey Rockwell and Stéfan Sinclair's *Hermeneutica*; and the many ongoing contributions to the journal *Cultural Analytics*.
19. Emily Apter, "Global Translatio: The 'Invention' of Comparative Literature, Istanbul, 1933," *The Translation Zone: A New Comparative Literature* (Princeton, NJ: Princeton University Press, 2005), 41–64.
20. Stephen Greenblatt and Catherine Gallagher, *Practicing New Historicism* (Chicago: University of Chicago Press, 2000), 6.
21. For an introduction, see Andrew Piper, "Think Small: On Literary Modeling," *PMLA* (forthcoming).
22. Hans Vaihinger, *The Philosophy of "As If": A System of the Theoretical, Practical and Religious Fictions of Mankind* (London: Routledge & K. Paul, 1965).
23. As Arthur Fine writes in his work resurrecting Vaihinger, "He finds no realm of human activities, even the most serious of them, into which play and imagination fail to enter. These faculties are part of the way we think ('constructively'), approach social and intellectual problems ('imaginatively'), employ metaphor and analogy in our language, and relate to others." Arthur Fine, "Fictionalism," *Midwest Studies in Philosophy* 18 (1993): 1–18.
24. On transcendence and computing culture, see Erik Davis, *TechGnosis: Myth, Magic, and Mysticism in the Age of Information* (New York: Harmony, 1998). On the latent theologism of literary studies, see Chad Wellmon, "Sacred Reading from Augustine to the Digital Humanists," *Hedgehog Review* 17.2 (Fall 2015): 70–84.
25. For a critique of the emphasis on size at the expense of "scale," of the integration of large and small, see English and Underwood, "Shifting Scales." For a thoughtful piece on algorithms as forms of representation, see Benjamin Schmidt, "Do Digital Humanists Need to Understand Algorithms?," *Debates in Digital Humanities* (2016), http://dhdebates.gc.cuny.edu/debates/text/99.
26. For an important reflection on the nature of literary data and historical representation, see Katherine Bode, "The Equivalence of 'Close' and 'Distant' Reading; or, Toward a New Object for Data-Rich Literary History," *MLQ* 78.1 (2017): 77–106.
27. For the foundational work in this direction, see Michael Polanyi, *The Tacit Dimension* (Chicago: University of Chicago Press, [1966] 2009). More re-

cently, see Gabriele Contessa, "Scientific Representation, Interpretation and Surrogative Reasoning," *Philosophy of Science* 74 (2007): 48–68.
28. On the problem of closed algorithms, see Cathy O'Neil, *Weapons of Math Destruction: How Big Data Increases Inequality and Threatens Democracy* (New York: Crown, 2016).
29. Deleuze, *Foucault*, 30–31, my emphasis.
30. Bruno Latour, *On the Modern Cult of the Factish Gods*, trans. Catherine Porter (Durham, NC: Duke University Press, 2010), 61.
31. Roland Barthes, *The Pleasure of the Text*, trans. Richard Miller (New York: Hill and Wang, 1975), 49, emphasis in original. For another critique of the sentence as the organizing principle of literary criticism, see M. M. Bakhtin, "The Problem of Speech Genres," *Speech Genres and Other Late Essays,* trans. Vern W. McGee (Austin: University of Texas Press, 1986), 71.
32. For an excellent introduction, see Peter D. Turney and Patrick Pantel, "From Frequency to Meaning: Vector Space Models of Semantics," *Journal of Artificial Intelligence Research* 37 (2010): 141–188. On the probabilistic nature of language acquisition and use more generally, see Rens Bod, Jennifer Hay, and Stefanie Jannedy, eds., *Probabilistic Linguistics* (Cambridge, MA: MIT Press, 2003).
33. Such a theory of meaning is by no means alien to the field of literary criticism. Hans Robert Jauss's theory of an aesthetic "horizon of expectation [*Erwartungshorizont*]" is fundamentally probabilistic in nature. The value of Jauss's model is the way it does not posit a universal sense of meaning, one that is true for all time. Rather, meaning is always highly situated. Within *this* culturally situated horizon of expectation, we see *this* degree of deviation. Norms, like their deviations, are always relative. Hans Robert Jauss, "Literary History as a Challenge to Literary Theory," *Toward an Aesthetic of Reception*, trans. Timothy Bahti (Minneapolis: University of Minnesota Press, 1982).
34. See Daniel Kahneman, *Thinking, Fast and Slow* (New York: FSG, 2013), 51.
35. For an introduction to understanding vector space models, see Turney and Pantel, "From Frequency to Meaning." Jockers also has a nice description in *Macroanalysis*, 63–67.
36. I recommend the Wikipedia entry for an initial understanding of all measures used throughout this book. Given how foreign these methods are to our discipline, a handbook geared toward humanists is certainly in order. In this particular case, see https://en.wikipedia.org/wiki/Cosine_similarity.
37. Roland Barthes, "Death of the Author," *Image—Music—Text*, trans. Stephen Heath (New York: Hill and Wang, 1978), 146.
38. J. Hillis Miller, "The Sense of an Un-Ending: The Resistance to Narrative Closure in Kafka's *Das Schloss*," *Franz Kafka: Narration, Rhetoric, Reading*, ed. Jakob Lothe, Beatrice Sandberg, and Ronald Speirs (Columbus: Ohio State University Press, 2011), 108–122, 114. Miller continues, "*The Castle,* no reader can doubt, is a potentially endless variation on the theme of failed 'attempts at passage'" (120).

39. "Of course, someone might object that my claim that no globally unified reading is possible is a globally unified reading." Miller, "The Sense of an Un-Ending," 111.
40. The data used for this experiment is drawn from the txtLAB450 trilingual collection of novels in German, French, and English published between 1770 and 1930 (NOVEL450). Only German novels were included in this test. In order to calculate "distinctive" words, I use both a Fisher's exact test and a G-test to calculate the log-likelihood ratio. When I speak of distinctive features, these will be the primary two tests I will use throughout this book. I use both because each measure provides a different perspective on semantic distinctiveness, with the Fisher's odds ratio favoring words that occur less frequently (and thus are potentially more semantically unique), where the G-test favors words that occur more frequently and are thus more prominently represented in the text. Like all things in this book, "distinctiveness" depends on the way it is measured. According to these tests, "Weg" is 47% more likely to appear in Kafka's *Castle* than in German novels generally. However, the p-value ($p = 0.002$), which indicates the likelihood that this is not an effect of chance, does not pass a statistical significance test. While throughout this book I will be using $p = 0.05$ as a standard threshold of significance (meaning the result is expected to occur in less than 95% of all cases), in this case, given that we are testing over 3,000 variables, we need to use what is known as a Bonferroni correction, which adjusts the p-value depending on the number of hypotheses being tested. Here we consider every word to be an individual hypothesis, so the better p-value would be 0.05/3000, or 0.000016. The log-likelihood ratio of 8.9 also indicates that the significance of this term is very weak. For more elaboration, see script 0.3.
41. For an introduction, see the aforementioned article Turney and Pantel, "From Frequency to Meaning."
42. The following steps were used to arrive at these insights. First, all words that co-occur within +/–9 words were counted for every novel in the collection, where stopwords were removed before the window was established (meaning it takes 9 non-stop words in either direction of *Weg*). This is the local "context" of *Weg*. These counts are then aggregated into a single table, where each individual novel's counts are subtracted from the total vector for all novels (so that when comparing a novel to the general representation of *Weg*, the novel itself is not included in all novels). Finally, each novel is then compared to the general semantic representation of *Weg* using cosine similarity. Because there is a very strong correlation between the similarity of a novel and the number of occurrences of *Weg* within the novel, I normalize similarity by dividing it by the log of the count of *Weg*. With this adjustment, the correlation between the *Weg*-count and similarity score drops from 0.83 to 0.26. After normalization, Kafka's similarity is only 0.64 standard deviations above the mean similarity score for all novels. See script 0.4.

43. These words are all ranked within the top 30 most distinctive words according to log-likelihood ratios. See script 0.3 and the table MDW_Kafka.csv for the list of words.
44. As Miller writes, "Unverifiable and inconsistent 'exegesis' forms the basic stylistic fabric of the novel." Miller, "The Sense of an Un-Ending," 113. For computational work in this direction, see J. Berenike Herrmann, "'Läuse im Pelz der Sprache?' Zu Funktionen von Modalpartikeln in narrativen (De-) Motivierungsstrategien bei Franz Kafka," *Die biologisch-kognitiven Grundlagen narrativer Motivierung*, ed. M. Horváth and K. Mellmann (Münster: Mentis, 2016).
45. Here the work of Rita Felski once again provides useful theoretical orientation. See Rita Felski, "Context Stinks!," *New Literary History* 42.4 (2011): 573–591.
46. Gertrude Stein, *The Autobiography of Alice B. Toklas* (New York: Vintage, 1990), 41.
47. Khan's work is part of a larger series of photographs of superimposed canonical works profiled in Garrett Stewart, *Bookwork: Medium to Object to Concept to Art* (Chicago: University of Chicago Press, 2011), 224–225.
48. Sybille Krämer, "*Schriftbildlichkeit*, oder: Über eine (fast) vergessene Dimension der Schrift," *Bild, Schrift, Zahl*, ed. Sybille Krämer and Horst Bredekamp (Munich: Fink, 2009), 157–176; Garrett Stewart, *The Look of Reading: Book, Painting, Text* (Chicago: University of Chicago, 2006); and Andrew Piper, "Face, Book," in *Book Was There: Reading in Electronic Times* (Chicago: University of Chicago, 2012), 25–44.
49. The recourse to the diagram belongs to the longer history of the spatialization of knowledge, from the medieval tree of knowledge, to the early modern schemas of Petrus Ramus and his school, to the eighteenth-century work of William Playfair and Johann Lambert, to the nineteenth-century diagrammatic theories of C. S. Peirce and John Venn, down to today's growing interest in the field of information visualization. For recent work on the history of the diagram, see John Bender and Michael Marrinan, *The Culture of the Diagram* (Stanford, CA: Stanford University Press, 2010); and Matthias Bauer and Christopher Ernst, *Diagrammatik: Einführung in ein kultur- und medienwissenschaftliches Forschungsfeld* (Bielefeld: Transcript, 2010). For a review of current information visualization practices, see Manuel Lima, *Visual Complexity: Mapping Patterns of Information* (Princeton, NJ: Princeton Architectural Press, 2011); and the new work of C. Foster, A. Hayward, S. Kucheryavykh, S. Negi, and L. Klein, "The Shape of History: Reimagining Nineteenth-Century Data Visualization," *Digital Humanities 2017* (Association of Digital Humanities Organizations, 2017).

CHAPTER ONE

1. The standard history is M. B. Parkes, *Pause and Effect: An Introduction to the History of Punctuation in the West* (Berkeley: University of California Press,

1993). In German, Stefan Höchli, *Zur Geschichte der Interpunktion im Deutschen* (Berlin: De Gruyter, 1981).
2. On the history of grammars, see Cecilia Watson, "Points of Contention: Rethinking the Past, Present, and Future of Punctuation," *Critical Inquiry* 38 (Spring 2012): 649–672.
3. These numbers are derived using the NOVEL450 data set. Results are contained in Punctuation_Novels_450.csv. See scripts 1.1–1.2.
4. The answer is not more dialogue or a sudden abbreviation of a name, as in "K." And yet the change is so dramatic, it suggests a major stylistic shift, even authorship. More could be explored here.
5. On the increasing alignment of writing with speech, see Naomi Baron, *Always On* (Oxford: Oxford University Press, 2008).
6. This finding is derived by calculating the number of periods per 1,000 words of text and then fitting a linear model to the values. Positive values indicate an increase in periods over the course of the novel, while negative values indicate a decrease in periods. The ratio of increase in periods to decrease in 19C novels is 1.52 (NOVEL 750), while for the twentieth century it is 1.86 (NOVEL_20C). See script 1.3.
7. Georges Bataille, *The Accursed Share*, vol. 1, trans. Robert Hurley (New York: Zone, 1991).
8. Wilhelm Lexis, *Zur Theorie der Massenerscheinungen in der menschlichen Gesellschaft* (Freiburg: Wagner, 1877), 1.
9. F. T. Marinetti, "Supplement to the Technical Manifesto," *Critical Writings*, ed. Günter Berghaus (New York: Farrar, Straus & Giroux, 2006).
10. See script 1.4 for this calculation.
11. All of the calculations that follow are based on the table Punctuation_Poetry_20C_All.csv, which was produced using script 1.0.
12. A word's increased likelihood of appearing at the end of a sentence was calculated using a Fisher's exact test. Only words that appeared more than 1,500 times or in approximately 2% of all documents were kept. This removes words that appear very infrequently but may occur more frequently at the end of a sentence. See script 1.5.
13. Paul Muldoon, *The End of the Poem* (New York: FSG, 2006).
14. A Fisher's exact test was used to measure the increased/decreased likelihood that a feature would occur in the high-period group. To avoid words that appear very infrequently, I kept only those words that appear more than 92 times, that is, in approximately 10% of all documents. See script 1.6.
15. The full results are contained in all tables labeled "Distinctive_."
16. Word length was calculated by counting the average number of characters per word. This is a standard measure of vocabulary difficulty used in numerous "reading ease" tests. Using a Wilcoxon rank-sum test, we see no significant difference between the median value of characters per word in the high-period group (M = 4.307) and the general population of poems (M = 4.326), W = 4,552,100, p = 0.9311. See script 1.7.

17. Negation was measured using the lexicon "negation," drawn from the Linguistic Inquiry Word Count Software (LIWC). See chapter four for a fuller discussion of this resource.
18. For a useful discussion of word embeddings for semantic analysis, see Ben Schmidt, "Vector Space for the Digital Humanities" (October 25, 2015), http://bookworm.benschmidt.org/posts/2015-10-25-Word-Embeddings.html.
19. For an implementation of the model, see script 1.8.
20. This claim is based on five separate trials using similarly sized random samples compared to the high-period corpus. A Wilcoxon rank-sum test shows that the median standard deviation is significantly higher for high-period poems in all trials:

Trial 1 W = 257640, p-value = 1.96e-05
Trial 2 W = 258610, p-value = 1.06e-05
Trial 3 W = 251860, p-value = 0.00054
Trial 4 W = 259450, p-value = 6.14e-06
Trial 5 W = 261210, p-value = 1.86e-06

However, using Cohen's d as a measure of effect size, we see that the effect of this difference is small (.20–.27).

CHAPTER TWO

1. For a fabulous exception that looks for the phonetic bridges between words—that reads across spaces—see Garrett Stewart, *Reading Voices: Literature and the Phonotext* (Berkeley: University of California Press, 1990).
2. While the keywords approach is so widespread as to resist any single citation, a nice example is represented by Brad Pasanek, *Metaphors of Mind: An Eighteenth-Century Dictionary* (Baltimore: Johns Hopkins University Press, 2015).
3. Rens Bod, Jennifer Hay, and Stefanie Jannedy, *Probabilistic Linguistics* (Cambridge, MA: MIT Press, 2003).
4. See Matthew L. Jockers, "Revealing Sentiment and Plot Arcs with the Syuzhet Package" (February 2, 2015), http://www.matthewjockers.net/2015/02/02/syuzhet/. This represents a new direction from his earlier work on topic modeling novels in Matthew L. Jockers and David Mimno, "Significant Themes in 19th Century Literature," *Poetics* 41 (2013): 750–761; or the second chapter of Matthew L. Jockers and Jodie Archer, *The Bestseller Code* (New York: St. Martin's, 2016).
5. Gérard Genette, *Narrative Discourse: An Essay in Method* (Ithaca, NY: Cornell University Press, 1980), 26.
6. In addition to Jockers, see Ben M. Schmidt, "Plot Arceology: A Vector-Space Model of Narrative Structure," *2015 IEEE International Conference on Big Data* (October 2015), 1667–1672; and Andrew J. Reagan, Lewis Mitchell,

Dilan Kiley, Christopher M. Danforth, and Peter Sheridan Dodds, "The Emotional Arcs of Stories Are Dominated by Six Basic Shapes," doi:10.1140/epjds/s13688-016-0093-1.
7. Georg Lukács, *Die Theorie des Romans: Ein geschichts-philosophischer Versuch über die Formen der großen Epik* (Munich: DTV, [1963] 2000), 73.
8. On the novelty and duration of Augustine's model, see Georg Misch, *Geschichte der Autobiographie*, 4 vols. (Bern: Francke, 1949); James Olney, ed., *Autobiography: Essays Theoretical and Critical* (Princeton, NJ: Princeton University Press, 1980); Charles Taylor, *Sources of the Self: The Making of Modern Identity* (Cambridge, MA: Harvard University Press, 1989); Robert Folkenflik, ed., *The Culture of Autobiography: Constructions of Self-Representation* (Stanford: Stanford University Press, 1993); and Patrick Coleman, Jayne Lewis, and Jill Kowalik, eds., *Representations of the Self from the Renaissance to Romanticism* (Cambridge: Cambridge University Press, 2000).
9. On the foundational relationship between narrative and causation, see Genette, *Narrative Discourse*, 26; and Tzvetan Todorov, *Introduction to Poetics* (Minneapolis: Minnesota University Press, 1981), 41. In its investment in a strong sense of before and after, conversional narration violates Greimas's theory of narrative equilibrium as the foundation of all narration. See A. J. Greimas, "Narrative Grammar," *MLN* 86 (1971): 793–806. On the function of narrative as one for marking difference, see Todorov's later work, *Genres in Discourse* (Cambridge: Cambridge University Press, 1990), 30.
10. Each book of the *Confessions* is ingested as a separate document. A list of stopwords are removed, and only words that appear in 60% of documents are kept. Word counts are normalized by the length of each book (meaning they are represented as percentages of total words per book). The "distance" between books is calculated using a Euclidean distance measure. See script 2.1 for implementation.
11. Debates surrounding the integrity of Augustine's work are legion, specifically whether books 10–13, which appear the most dissimilar here, should be considered a "supplement" to the pre-conversional books or an integral part of the whole. The critical uncertainty that surrounds the unity of the *Confessions* is in many ways a reflection of precisely this aspect of discursive heterogeneity that belongs to the narrative of conversion and that the distributional model brings out. More importantly, the work has also *historically* only ever been mediated to readers as a whole. As James O'Donnell writes, "There is no evidence that the work ever circulated in a form other than the one we have." James O'Donnell, *The Confessions of Augustine: An Electronic Edition* (1992), http://www.stoa.org/hippo/comm.html. According to a scribal and printed tradition of reproduction, then, the *Confessions* is most often reproduced as a unity and thus historically understood as such. For debates on the integrity of the text, see in particular J. J. O'Meara, *The Young Augustine* (London, 1954); and Pierre Courcelle, *Les Confessions de Saint Augustin dans la tradition littéraire* (Paris, 1963).

12. O'Donnell, *The Confessions of Augustine*, http://www.stoa.org/hippo/comm 10.html#CB10C1S1.
13. See Andrew Piper, "Novel Devotions: Conversional Reading, Computational Modeling, and the Modern Novel," *New Literary History* 46.1 (Winter 2015): 63–98. This chapter presents a slightly revised method for detecting lexical contraction/expansion than in the original essay. See the next note for a fuller description of the method used.
14. The method used to derive these scores consisted of the following steps: works were divided into twenty equal parts (to approximate ideal "chapters" and control for different chapter lengths); only words were kept that appear in a majority of parts (i.e., greater than 10 sections); stopwords were removed; and words were stemmed using the SnowballC package in R. The idea that I am trying to model is expressive contraction, which I approximate through a measure of more similar vocabulary distributions between the first and second halves of a novel. In order to assess the degree of difference between the halves *relative to themselves*, for each half I create a vector that represents the similarities of each part of that half to every part of that half (similarity of part 1–part 2, similarity of part 1–part 3, etc., and then again for the second half, 11–12, 11–13, 11–14, etc.). The "contraction score" is calculated by taking the percentage decrease of the median distance of the second-half parts to each other relative to the median distance of the first-half parts to each other. Larger negative numbers indicate a greater amount of contraction, that is, there is more semantic similarity in the second half of the novel when compared to the first. See script 2.2. Results are located in Contraction_All_Trial1.csv.
15. All lists of distinctive words are located in the directory "MDW" and were generated using script 2.3.
16. In order to extract the social network of a novel, I use the following process. First, I input a list of proper names (this can be automated using named entity recognition). Second, I divide a given novel into 250 word units or "pages." For every page, I draw an edge between every proper name that co-occurs on that page (regardless of the number of appearances of the name—if "Lene" appears three times and "Botho" appears once, there will still only be 1 page edge between Lene–Botho). The number of edges for a given section are combined to create a single "edge weight" for any given relationship between two characters. The edge weight thus represents the number of page co-occurrences for any two names. For example, in the first quarter of the novel, Botho and Käthe appear on the same page 2 separate times, so they have an edge weight of 2 (Botho and Lene appear, by contrast, 12 times together). In the final quarter of the novel, Botho and Käthe now appear 16 separate times together, the most of any pair (Botho and Lene drop to half of that, at 7). See script 2.4.
17. Friedrich Betz, *Theodor Fontane*, Irrungen, Wirrungen*: Erläuterungen und Dokumente* (Stuttgart: Reclam, 1979), 86.

18. On the novel's polyphony, see Ingrid Mittenzwei, *Die Sprache als Thema: Untersuchungen zu Fontanes Gesellschaftsromanen* (Bad Homburg: Gehlen, 1970); Horst Schmidt-Brümmer, *Formen des perspektivischen Erzählens: Fontanes Irrungen, Wirrungen* (München: Fink, 1971); Norbert Mecklenburg, *Theodor Fontane: Romankunst der Vielstimmigkeit* (Frankfurt/Main: Suhrkamp, 1998); and most recently Gerhard Neumann, *Theodor Fontane: Romankunst als Gespräch* (Freiburg: Rombach, 2011).
19. Jonathan Crary, *Suspensions of Perception: Attention, Spectacle, and Modern Culture* (Cambridge, MA: MIT Press, 1999).
20. For an example of TTR in action, see Ian Lancashire and Graeme Hirst, "Vocabulary Changes in Agatha Christie's Mysteries as an Indication of Dementia: A Case Study," presented at the 19th Annual Rotman Research Institute Conference, *Cognitive Aging: Research and Practice*, March 8–10, 2009.
21. I use a subset of 1,000 novels published between 1880 and 1930 randomly selected from the Chicago Text Lab corpus (MODERN_20C). In order to calculate vocabulary richness, I take 100 random samples of 1,000-word windows for every novel and average the TTR score across all runs. The reason for this approach is to control for differing lengths of novels (TTR is extremely sensitive to text length). By doing so, the correlation between length and TTR in my model drops to $r = 0.027$, $p = 0.3943$. See script 2.5.
22. Joyce_TTR_Synonymy.csv. See script 2.6.
23. I derive synonymy scores for a given page by using the resource WordNet. Synonymy is calculated as the percentage of words that appear in the synset of at least one other word on the page. It is based on the premise that if a word appears in the list of synonyms for another word, then those words can be considered to be synonyms of one another. See the script calc_synonyms_antonyms.py.
24. See script 2.7.
25. Output is found in Contraction_All_Tria12.csv. See script 2.8.
26. All outputs are contained in the MDW folder. See script 2.9.
27. Results are based on calculating the number of proper names for sliding 1,000-word windows (across 100-word increments). A Wilcoxon signed-rank-sum test shows that the second half of the novel contains significantly more proper names per window (W = 28,866, $p < 2.2e$-16). The median value of proper names in a window in the first half is 3.96% of all tokens, while in the second half it is 5.66%. This translates into an average increase of 8–9 more proper names per page. See script 2.10.
28. Koustuv Sinha, Alayne Moody, Derek Ruths, and Andrew Piper, "On the Unreasonable Complexity of Detecting Social Interactions in Literature" (forthcoming). For an example of the use of social networks to study literature, see Graham Sack, "Character Networks for Narrative Generation: Structural Balance Theory and the Emergence of Proto-Narratives," *Workshop on Computational Models of Narrative* (Dagstuhl: Dagstuhl Publishing, 2013); Mariona Coll Ardanuy and Caroline Sporleder, "Structure-Based

Clustering of Novels," *Proceedings of the 3rd Workshop on Computational Linguistics for Literature (CLfL) @ EACL 2014*, 31–39, Gothenburg, Sweden, April 27, 2014; and Prashant Arun Jayannavar, Apoorv Agarwal, Melody Ju, and Owen Rambow, "Validating Literary Theories Using Automatic Social Network Extraction," *Proceedings of NAACL-HLT Fourth Workshop on Computational Linguistics for Literature*, 32–41, Denver, Colorado, June 4, 2015.

29. Alex Woloch, *The One vs. the Many: Minor Characters and the Space of the Protagonist in the Novel* (Princeton, NJ: Princeton University Press, 2003).
30. See script 2.11. Jonathan Foote, "Automatic Audio Segmentation Using a Measure of Audio Novelty," *Proceedings of IEEE International Conference on Multimedia and Expo*, vol. 1 (2000): 452–455. For an application of this technique to understanding changes in musical style in popular music, see Matthias Mauch, Robert M. MacCallum, Mark Levy, and Armand M. Leroi, "The Evolution of Popular Music: USA 1960–2010," *Royal Society Open Science* (2015), doi:10.1098/rsos.150081. For a response, see Ted Underwood, Hoyt Long, Richard Jean So, and Yuancheng Zhu, "You Say You Found a Revolution," *The Stone and the Shell* (2016), https://tedunderwood.com/2016/02/07/you-say-you-found-a-revolution/.
31. See Schmidt, "Plot Arceology"; Reagan et al., "Emotional Arcs"; and Jockers, "Revealing Sentiment."
32. This method is more fully described in Eva Portelance and Andrew Piper, "Narrative Frame Detection" (forthcoming).

CHAPTER THREE

1. Ernst Robert Curtius, "Begriff einer historischen Topik," *Zeitschrift für Romanische Philologie* 58 (1938): 129–142. Reprinted in *Toposforschung*, ed. Max L. Maeumer (Darmstadt: Wissenschaftliche Buchgesellschaft, 1973), 14, 3.
2. For an introduction to topic modeling to literature and culture, see the special issue of *Poetics*, "Topic Models and the Cultural Sciences," ed. John W. Mohr and Petko Bogdanov, 41.6 (2013): 545–770.
3. Xuerui Wang and Andrew McCallum, "Topics over Time: A Non-Markov Continuous-Time Model of Topical Trends," *Proceedings of the 12th ACM SIGKDD international conference on Knowledge discovery and data mining* (2006): 424–433; Jordan Boyd-Graber and David Blei, "Multilingual Topic Models for Unaligned Text," *Proceedings of the Twenty-Fifth Conference on Uncertainty in Artificial Intelligence* (2009): 75–82.
4. Jockers and Mimno, "Significant Themes in 19th Century Literature"; Matthew L. Jockers, "Theme," *Macroanalysis: Digital Methods and Literary History* (Chicago: University of Illinois Press, 2013), 119–153; Andrew Goldstone and Ted Underwood, "The Quiet Transformations of Literary Studies: What Thirteen Thousand Scholars Could Tell Us," *New Literary History* 45.3 (2014): 359–384; and Matt Erlin, "The Location of Literary History: Topic Modeling, Network Analysis and the German Novel, 1731–1864," *Distant Readings:*

Topologies of German Culture in the Long Nineteenth Century, ed. Matt Erlin and Lynne Tatlock (New York: Boydell and Brewer, 2014), 55–90.

5. Ann Moss, *Printed Commonplace-Books and the Structuring of Renaissance Thought* (Oxford: Clarendon, 1996).
6. Peter de Bolla, *The Architecture of Concepts: The Historical Formation of Human Rights* (New York: Fordham University Press, 2013). For an exploration of topics as a continuation of historical indexing practices in books, see Jeffrey M. Binder and Collin Jennings, "Visibility and Meaning in Topic Models and 18th-Century Subject Indexes," *Literary and Linguistic Computing* 29.3 (2014): 405–411. See also the new tool produced by JSTOR labs, the topicgraph, which visualizes topics across scholarly books: https://labs.jstor.org/topicgraph/.
7. Michel Foucault, *The Order of Things: An Archaeology of the Human Sciences* (New York: Vintage, [1970] 1994).
8. The work of Caroline Levine on forms and their foundations within literary criticism is an excellent framework through which to approach these issues. Caroline Levine, *Forms: Whole, Rhythm, Hierarchy, Network* (Princeton, NJ: Princeton University Press, 2015).
9. See Aristotle, *Topica*, trans. E. S. Forster, Loeb Classical Library (Cambridge, MA: Harvard University Press, 1989).
10. Aristotle, *Topica*, 163b28–32.
11. Moss, *Printed Commonplace-Books*.
12. J. R. McNally, "Rudolph Agricola's *De inventione dialectica libri tres*: A translation of selected chapters," *Speech Monographs*, 34.4 (1967): 393–422, doi:10.1080/03637756709375551.
13. Cited in Moss, *Printed Commonplace-Books*, 224.
14. Erasmus, *De copia*: *The Collected Works of Erasmus*, vol. 2, ed. Craig Thompson (Toronto: University of Toronto Press, 2016), 638.
15. Erasmus, *De copia*, 636.
16. On the continuation of commonplacing, see Stephen Colclough, "Recovering the Reader: Commonplace Books and Diaries as Sources of Reading Experience," *Publishing History* 44 (1998): 5–37. For differing accounts as to the causes of commonplacing's demise, see Ann Moss and Mary Thomas Crane, *Framing Authority: Sayings, Self, and Society in Sixteenth-Century England* (Princeton, NJ: Princeton University Press, 2014).
17. As Ann Moss points out, the critique of "la pedanterie des Compilateurs" had already been a common theme in the seventeenth century. The phrase is from Jean-Louis Guez de Balzac's *Paraphrase, ou la grande eloquence* (1640). See Moss, *Printed Commonplace-Books*, 259. On the rise of paraphrastic thought via the work of Goethe, see Andrew Piper, "Paraphrasis: Goethe, the Novella, and Forms of Translational Thought," *Goethe Yearbook* 17 (2010): 179–201.
18. *Encyclopédie ou Dictionnaire raisonné des sciences, des arts et des métiers, par une Société de Gens de lettres*, vol. 1, 361.

19. Brad Pasanek and Chad Wellmon, "The Enlightenment Index," *The Eighteenth Century: Theory and Interpretation* 56:3 (2015): 357–380.
20. Moss, *Printed Commonplace-Books*, 281.
21. The mode was generated using the "topicmodels" package in R. The method of Gibbs sampling was used with an alpha setting of 0.5. See script 3.1 for the model parameters, and for the output see the table 19C_NovelPlusPoetry _Words_Topic100_Alpha0.5.csv.
22. Deleuze, *Foucault*, 55.
23. The model was generated using the "topicmodeling" package in R and applying the VEM method to 1,000-word passages. Three hundred topics were chosen. Words were stemmed, and an extended stopword list including proper names was applied. The model was applied to the German novels of the NOVEL450 data set. See script 3.2.
24. Model "optimization" or model "fit" is a long-standing research challenge in the field. On model optimization based on semantic coherence, see David Mimno, Hanna Wallach, Edmund Talley, Miriam Leenders, and Andrew McCallum, "Optimizing Semantic Coherence in Topic Models," *Proceedings of the 2011 Conference on Empirical Methods in Natural Language Processing* (2011): 262–272.
25. Bruno Latour, "Why Has Critique Run Out of Steam? From Matters of Fact to Matters of Concern," *Critical Inquiry* 30.2 (2004): 151–174.
26. See script 3.2.
27. For a discussion, see Erlin, "The Location of Literary History."
28. Heterogeneity is essentially an inverse purity score. One could also use an entropy score to capture the overall balance of passages to novels for a given topic. Here, I am more interested in capturing the extent to which a single novel dominates a topic.
29. For a discussion of semantic coherence in topic models, see Mimno et al., "Optimizing Semantic Coherence."
30. Tf-idf refers to "term frequency—inverse document frequency." It is a technique of normalizing word counts in documents where the term's frequency (the number of times it appears in the document) is divided by the number of documents in which that word appears. The more documents a word appears in, the lower the score will be. This means that tf-idf privileges words that are important to a document but rarer across a corpus (i.e., more distinctive of that document compared to the rest of the corpus).
31. All citations that follow are drawn from the NOVEL450 data set. The citations refer to individual files that represent 1,000-word passages and can be found in the data accompanying this chapter. All translations are my own. This passage can be found in 1774_Goethe,Johann_DieLeidendesjungen-Werthers1_Novel_24000.txt.
32. Cited from 1799_Hölderlin_Hyperion_Novel_21000.txt.
33. Cited from 1796_Tieck,Ludwig_WilliamLovell_Novel_20000.txt.
34. See 1802_Brentano,Clemens_Godwi_Novel_22000.txt.

35. On fluidity and Romanticism, see David Wellbery, "Primordial Song," *The Specular Moment: Goethe's Early Lyric and the Beginnings of Romanticism* (Stanford, CA: Stanford University Press, 1996), 233–284.
36. Surprisingly little has been written on the conjunctions between *Hyperion* and *Werther* (seven articles concern both novels out of a total of 651 on *Werther*, according to the *Bibliographie der deutschen Sprach- und Literaturwissenschaft*). Interestingly, *Hyperion* is the seventh most similar novel to *Werther* out of 150 German novels (*William Lovell* and *Godwi* are two of the others). See also Doris Feil's monograph on the different epistemological aims of the two novels. Doris Feil, *Stufen der Seele: Erkenntnistheoretische Darstellung in Goethes "Werther" und Hölderlins "Hyperion"* (Oberhausen: Athena, 2005).
37. Cited in 1799_Hölderlin_Hyperion_Novel_16000.txt.
38. The three passages are 1910_Rilke,RainerMaria_MalteLauridsBrigge_Novel_17000.txt, 1774_Goethe,Johann_DieLeidendesjungenWerthers1_Novel_24000.txt, and 1799_Hölderlin_Hyperion_Novel_21000.txt. For the clustering, I use Ward's hierarchical clustering algorithm based on cosine similarity between documents for the top twenty topic words. See script 3.3.
39. Ulrich Fülleborn, "'Werther'—'Hyperion'—'Malte Laurids Brigge': Prosalyrik und Roman," *Studien zur Deutschen Literatur: Festschrift für Adolf Beck zum siebzigsten Geburtstag*, ed. Ulrich Fülleborn and Johannes Krogoll (Heidelberg: Winter, 1979).
40. Here I rely on Stephen Mitchell's translation to help with this dizzying passage. Rainer Marie Rilke, *The Notebooks of Malte Laurids Brigge*, trans. Stephen Mitchell (New York: Vintage, 1985), 76.
41. As I have written elsewhere regarding Werther's circulation within Goethe's own writing, the jug as a metaphor for art engages in a wider debate about the shifting understanding of art no longer understood as a form of *Handwerk* or handicraft in the eighteenth century. See Andrew Piper and Mark Algee-Hewitt, "The Werther Effect I: Goethe, Objecthood, and the Handling of Knowledge," *Distant Readings: Topologies of German Culture in the Long Nineteenth Century*, ed. Matt Erlin and Lynn Tatlock (Rochester, NY: Camden House, 2014), 155–184.
42. Gaston Bachelard, *Water and Dreams: An Essay on the Imagination of Matter*, trans. Edith R. Farrell (Dallas: Dallas Institute of Humanities and Culture, 1983), 6.
43. Georges Bataille, citing Alexandre Kojève, remarks that Hegel's philosophy, which he sees as nothing more than a transcoding of *Werther*, is a philosophy of death. Seen in this light, Romanticism is the articulation of a subjectivity as a form of conditional dying, to which only art gives full access. See George Bataille, "Hegel, Death, and Sacrifice," trans. Jonathan Strauss, *Yale French Studies* 78 (1990): 9–28.
44. See script 3.4 for implementation.

45. Michel Foucault, *The Order of Things*, 313. On Romanticism and the rise of anthropology, see Maureen N. McLane, *Romanticism and the Human Sciences: Poetry, Population, and the Discourse of the Species* (Cambridge: Cambridge University Press, 2000); Stefan Schweizer, *Anthropologie der Romantik: Körper, Seele, Geist* (Paderborn: Schöningh, 2008); and Albrecht Koschorke, *Körperströme und Schriftverkehr: Mediologie des 18. Jahrhunderts* (Stuttgart: Fink, 1999).
46. The steps to do so are the following. First, I create a table of the frequencies of the top twenty topic words from Topic 150 for the thirty documents in which this topic is strongly present. Instead of representing our documents as vectors of word frequencies, as I did with Augustine in the previous chapter, here I am representing the terms as vectors of frequencies within documents. This allows me to observe the similarity of usage across documents. Using a measure of "correlation," I then transform this term-document matrix into a similarity matrix, where every word has a similarity value (correlation) to every other word. The more similar the relative frequencies are across the documents, the more two words will be correlated. Here is a simple example of three terms, with frequencies in ten documents:

Document Frequencies	
Term1	0,2,4,0,7,3,0,8,2,0
Term2	0,1,3,0,3,5,1,4,0,0
Term3	3,0,0,4,1,2,5,0,6,2

This would give a correlation of 0.75 for Term1 and Term2 and –0.66 for Term1 and Term3. You can see how Term3 rises where Term1 and Term2 typically fall and vice versa.

Doing this on our data yields a 20x20 matrix where every word has a correlation value to every other word. This is then reduced to two dimensions using multidimensional scaling to represent as a graph. Another approach to understanding these relationships would be to use the word embeddings model that I discussed in chapter one, which would generate relationships between these keywords based on their semantic contexts (words they have in common nearby). I chose not to go this route for two reasons: 1) with so little text, these kinds of models are very unstable (it generates very different outputs for every run of the model); and 2) I am interested in how these terms correlate with each other in the documents under review rather than understanding their overall semantic contexts. How *life* and *death* mirror each other in their frequencies is closer to my question than how their semantic contexts mirror each other (or don't). See script 3.5.
47. Umberto Eco, *Semiotics and the Philosophy of Language* (Bloomington: Indiana University Press, 1984), 68.
48. See script 3.6.

NOTES TO PAGES 87–97

49. For a list of all documents that are weakly associated with Topic 150, see the table Topic150_Document_List_Weak_State.csv.
50. Cited from 1803_Mereau,Sophie_AmandaundEduard_Novel_4000.txt.

CHAPTER FOUR

1. For a thorough discussion, see Käthe Hamburger, *The Logic of Literature*, 2nd ed., trans. Marilyn J. Rose (Bloomington: Indiana University Press, 1973), 233; and Gérard Genette, *Fiction and Diction*, trans. Catherine Porter (Ithaca, NY: Cornell University Press, 1993), 1–29.
2. These features are drawn from the Linguistic Inquiry and Word Count Software (LIWC 2007), which will be discussed in more detail later in the chapter. For a discussion, see Yla R. Tausczik and James W. Pennebaker, "LIWC and Computerized Text Analysis Methods," *Journal of Language and Social Psychology* 29.1 (2010): 24–54.
3. Initial successful work has been done on the predictability of fiction by Ted Underwood and that of different novel genres by Matthew Jockers. The aim of this chapter is to begin to identify and interpret the particular features that are indicative of fictional writing and what those features have to tell us about the nature of fictionality and the history of the novel in particular. See Ted Underwood, "Understanding Genre in a Collection of a Million Volumes, Interim Report," *Figshare*, https://dx.doi.org/10.6084/m9.figshare.1281251.v1; and Matthew L. Jockers, "Style," in *Macroanalysis: Digital Methods and Literary History* (Chicago: University of Illinois Press, 2013), 54–101.
4. Richard M. Gale, "The Fictive Use of Language," *Philosophy* 66 (1971): 324–340; David Lewis, "Truth in Fiction," *American Philosophical Quarterly* 15 (1978): 37–46; John R. Searle, "The Logical Status of Fictional Discourse," *Expression and Meaning: Studies in the Theory of Speech Acts* (Cambridge: Cambridge University Press, 1979), 58–76; Hilary Putnam, "Is There a Fact of Matter about Fiction?," *Poetics Today* 4.1 (1983): 77–82; Benjamin Hrushovski, "Fictionality and Fields of Reference," *Poetics Today* 5.2 (1984): 227–251; and Gregory Currie, *The Nature of Fiction* (Cambridge: Cambridge University Press, 1990).
5. John R. Searle, "Logical Status," 68. As we will see, everything depends on Searle's distinction here on "stretch of discourse," itself a stretch of discourse.
6. Maurice Blanchot, *The Instant of My Death*; and Jacques Derrida, *Demeure: Fiction and Testimony* (Stanford, CA: Stanford University Press, 2000), 28.
7. Stanley Fish, *Is There a Text in This Class? The Authority of Interpretive Communities* (Cambridge, MA: Harvard University Press, 1982).
8. Hardly a thing of the past, this position is now being replayed in the field of "postclassical" narratology, which argues that there are no inherent distinguishing features between fictional and nonfictional narratives.

Driven largely by new work in the theory of mind, attention is paid not to the unique features of texts but to the cognitive apparatus that is brought to bear on these texts and that is assumed to be common across all kinds of narration. As I will show, not only are fictional narratives highly distinct from nonfictional ones from a linguistic perspective, but their differences are most strongly driven by an attention to sense perception, that is, to a sense of embodiment, making a strict reliance on cognition as narrative's primary framework problematic. For the postclassical narratological position, see J. Alber and M. Fludernik, eds., *Postclassical Narratology* (Columbus: Ohio State University Press, 2010). This new work is driven largely as a response to the "classical" narratological work of Dorrit Cohn, *The Distinction of Fiction* (Baltimore: JHU Press, 1999); and even earlier, Käthe Hamburger, *The Logic of Literature* (Bloomington: Indiana University Press, 1973).

9. We await a study that compares the accuracy of machines with that of humans. My hypothesis is that they would mirror each other quite strongly but that humans would probably require less text—they would achieve higher levels of accuracy more quickly.

10. This graph was generated on the contemporary fiction data set and compared with works of contemporary history (NOVEL_CONT/NON_CONT). An SVM classifier was used with a Gaussian kernel using the kernlab package in R. The features used in the model are taken from the eighty features included in LIWC. "Accuracy" represents the average F1 score for tenfold cross-validation for each segment size. See script 4.1.

11. See the work of Ted Underwood that addresses precisely this idea of the stability of genre. Ted Underwood, "The Life Cycles of Genres," *Cultural Analytics* (May 23, 2016), http://culturalanalytics.org/2016/05/the-life-cycles-of-genres/. On the pre-nineteenth-century heterogeneity of fictional narrative, see Mark Algee Hewitt, Laura Eidem, Ryan Heuser, Anita Law, and Tanya Llewellyn, "Novel Taxonomies: Prehistories of Genre in the Eighteenth Century" (forthcoming). In a non-computational vein, Thomas Pavel argues that the unity of the nineteenth- and twentieth-century novel is an intentional aesthetic contrast to the early modern prioritization of "sub-genres." "Early modern narrative culture emphasized the differences between subgenres, while later forms of the novel are the result of multiple attempts to blend these subgenres together." Thomas Pavel, *The Lives of the Novel: A History* (Princeton, NJ: Princeton University Press, 2015), 10.

12. Cohn, *The Distinction of Fiction*, 25.

13. In this, I see my findings aligning closely with the "possible worlds theory" that was influential in narrative theory in the 1990s. See Thomas Pavel, *Fictional Worlds* (Cambridge, MA: Harvard University Press, 1989); Marie Laure-Ryan, *Possible Worlds, Artificial Intelligence, and Narrative Theory* (Bloomington: Indiana University Press, 1992); and Ruth Ronen, *Possible Worlds in Literary Theory* (Cambridge: Cambridge University Press, 1994). It

also lends support for the more recent new historical work of John Bender that emphasizes the relationship between the coeval rise of the novel in the eighteenth century and scientific experimentation. See John Bender, *Ends of Enlightenment* (Stanford, CA: Stanford University Press, 2012).

14. As Catherine Gallagher writes, "If a genre can be thought of as having an attitude, the novel has seemed ambivalent toward its fictionality—at once inventing it as an ontological ground and placing severe constraints upon it." See Catherine Gallagher, "The Rise of Fictionality," *The Novel*, vol. 1, ed. Franco Moretti (Princeton, NJ: Princeton University Press, 2006), 336–363; Elaine Freedgood, "Denotatively, Technically, Literally," *Representations* 125 (Winter, 2014): 1–14; Frances Ferguson, "Now It's Personal: D. A. Miller and Too-Close-Reading," *Critical Inquiry* 41 (Spring 2015): 527; and Ian Watt, *The Rise of the Novel: Studies in Defoe, Richardson, and Fielding* (Berkeley: California University Press, 2001). It should be added that Freedgood's emphasis is by no means normative in her identification of the importance of the novel's technical vocabulary; she simply wants us to attend to this overseen dimension because of the way it expands the archive of reading. As I will show at the close of this chapter, this attention to factuality within the novel is something that computational approaches are well suited to address. It is also worth noting how these positions all focus on an earlier time frame than I am exploring here, and yet their time frames still serve as the basis of normative arguments about the novel more generally ("our kind of fiction," in Michael McKeon's words). An open question is whether these ambiguities of fictionality or sense of heightened referentiality look that way because of what is happening in the seventeenth-century novel, that is, whether these arguments are based on a different kind of shift occurring around the turn of the eighteenth century that is no longer operative by the nineteenth.

15. Maurice Merleau-Ponty, *The Primacy of Perception*, ed. James M. Edie (Evanston, IL: Northwestern University Press, 1964), 17.

16. For an excellent discussion and application of predictive models, see Underwood, "The Life Cycles of Genres."

17. The novels in this collection are drawn from NOVEL450.

18. The designations of fiction and nonfiction are derived from Ted Underwood's collection, which is located at https://dx.doi.org/10.6084/m9.figshare.1281251.v1. All duplicate titles were removed; all documents with the word stems for "essays," "tales," "scenes," or "stories" in either the genre or title fields were removed; and only those works with an 89% or better chance of containing more than 80% pages of fiction were chosen.

19. Similar to the .txtLAB nineteenth-century collection, the documents here represent a canonical representation of fiction and nonfiction of the past ten years, meaning they have passed through some kind of filter, whether it is the *New York Times Book Review*, a literary prize competition short list, or various bestseller lists of platforms like Amazon.com or the *New York*

Times. For a more thorough review of the data set and its insights into contemporary forms of social value surrounding the novel, see Piper and Portelance, "How Cultural Capital Works."
20. LIWC 2007 was used in this chapter.
21. For those interested in studying the dictionaries used by LIWC, see their language manual at http://www.liwc.net/LIWC2007LanguageManual.pdf.
22. See Tausczik and Pennebaker, "LIWC and Computerized Text Analysis Methods."
23. For a thoughtful and practical review of lexicon-driven research, see H. A. Schwartz, J. E. Eichstaedt, L. Dziurzynski, M. L. Kern, E. Blanco, S. Ramones, M. E. P. Seligman, and L. H. Ungar, "Choosing the Right Words: Characterizing and Reducing Error of the Word Count Approach," *Proceedings of SEM-2013: Second Joint Conference on Lexical and Computational Semantics* (2013), Atlanta, GA, 296–305. For a comparison of lexicon versus machine-learning approaches to text analysis, see Piper and Portelance, "How Cultural Capital Works."
24. The kernlab package was used in R. All runs use a Gaussian kernel ("rbfdot"). See script 4.2 for further details. A useful introductory text to machine learning is Brett Lantz, *Machine Learning with R* (Birmingham, UK: Packt, 2013). For a discussion of SVMs in literary text classification, see Bei Yu, "An Evaluation of Text Classification Methods for Literary Study," *Literary and Linguistic Computing* 23.3 (2008): 327–343.
25. On the value of using a rank-sum test to test distributions of linguistic features, see Adam Kilgarriff, "Comparing Corpora," *International Journal of Corpus Linguistics* 6.1 (2001): 97–133.
26. See script 4.3 for implementation.
27. It is entirely possible to reverse this assumption and privilege features that are more prevalent overall, something that becomes valuable when assessing individual words. This is the approach I use in figure 4.6. Individual words can have such low frequencies that favoring words with higher counts ensures that one is finding more "important" or perhaps "relevant" vocabulary, that is, less random vocabulary. The crucial point is that once again the outcomes are determined by the initial assumptions used in the model.
28. All tables are included in the supplementary data.
29. See LIWC_Comparison_EN_FIC_v_HATHI_FIC.csv, located in the LIWC Feature Comparison Tables.
30. In addition to removing dialogue, 200 verbs of communication were removed from the texts and any personal pronouns that appeared within +/−1 word (I said, said she, etc.). For the purposes of visualization, exclamation marks have also been removed.
31. Using the NLP package in R, the mean number of unique named persons in a novel was 10.58 per 10,000 words, whereas it was 23.94 for history. See NER_19C_EN.csv and NER_19C_History.csv. See script 4.5.

32. When I test this question of sense perception using a more limited range of words curated by my lab, we see that overall nineteenth-century fiction is 2.5 times more likely to focus attention on the experience of sense perception (a level that I should add does not change by the twenty-first century). For the list of words used, see dict_sense_stem_all.csv and script 4.6.
33. The tables behind these figures are included in the supplementary data.
34. Lisa Zunshine, *Why We Read Fiction: Theory of Mind and the Novel* (Columbus: Ohio State University Press, 2006).
35. Underwood, "The Life Cycles of Genres."
36. Piper and Portelance, "How Cultural Capital Works."
37. This fiction/novel distinction that is translated across the axis of time will recapitulate itself in the genre distinctions of the contemporary novel as they relate to social value. One of the strongest ways bestselling and prize-winning novels in the present differentiate is across the feature of nostalgia and retrospection. See Piper and Portelance, "How Cultural Capital Works."
38. Results are contained in the table MDW_EN_NOV_3P_v_HATHI_TALES _Final.csv. See script 4.7.
39. This observation aligns with Matt Erlin's argument about the philosophical dimensions of novelistic narration. Matthew Erlin, "Topic Modeling, Epistemology and the English and German Novel," *Cultural Analytics* (April 26, 2017), doi:10.22148/16.014.
40. If we condition on just these four features, the classification results will outperform the average accuracy of four features chosen at random by a statistically significant margin, though to be sure the numbers are considerably lower than when we use all eighty features, and there are other combinations that will perform slightly better. Those feature combinations that do perform better most often contain present-tense verbs and categories of sense perception, pointing to the other ways discussed here through which novels are unique. All Features = 76%, Prevarication Features = 63.4%, Random4 (100 trials) = 60.8% +/−3.8%.
41. Underwood, "The Life Cycles of Genres."
42. This finding amplifies the work of Ryan Heuser and Long Le Khac in "A Quantitative Literary History of 2,958 Nineteenth-Century British Novels: The Semantic Cohort Method," *Stanford Literary Lab Pamphlet 4*, http://litlab.stanford.edu/LiteraryLabPamphlet4.pdf. As I will show in future work, however, there is a more nuanced story about the simple decline of abstraction/rise of concreteness narrative that they put forth.
43. Tracking either emotional states or sense perception is of course very challenging and will require much further research. For both, I use custom dictionaries curated by my lab that are designed to be more conservative than the LIWC lexica or emotion dictionaries such as the NRC lexicon. For example, the NRC lexicon contains words like *claw* or *abacus* that are labeled as emotions (anger and trust, respectively). The txtLAB lexicon contains a narrower range of words that are more explicitly indicative of

emotional states, such as *wonder, woe, qualm, phobia, misery, mirth*, and so on. While it will miss emotions that are present in a text that are metaphorically or more subtly expressed, it will do a better job of capturing the explicit mentions of emotions. The NRC and .txtLAB lists correlate in a medium way ($r = 0.44$, $r = 0.46$ for each category), which suggests they are capturing slightly different things. The .txtLAB emotion dictionary has only 453 words, while the NRC list has 6,468 unique words, and the .txtLAB perception dictionary consists of just 79 words. The data used in the tests combines the NOVEL_20C and NOVEL_19C data sets. See script 4.8. For a discussion of emotion detection, see Saif Mohammed, "From Once Upon a Time to Happily Ever After: Tracking Emotions in Mail and Books," *Decision Support Systems* 53 (2012): 730–741; and Saif Mohammed and Peter Turney, "Crowdsourcing a Word-Emotion Association Lexicon," *Computational Intelligence* 29.3 (2013): 436–465.

44. If we look more closely at the particular words driving this effect, we see that much of it can be explained by an increased vocabulary related to vision. It is one sense in particular that seems to be much more common in novelistic writing since the mid-twentieth century.

45. The data used in this experiment is drawn from the contemporary novels reviewed in the *New York Times* and 100 random samples drawn from the Stanford nineteenth-century collection. The counts are based on the LIWC categories of "body," "perception," and "affect." See script 4.9. Results are located in LIWC_Comparison_CONT_NOV_v_19C_Stanford.csv.

CHAPTER FIVE

1. Vladimir Propp, *Morphology of the Folktale*, trans. Laurence Scott (Austin: University of Texas Press, 1968).
2. For the most extended discussion, see Roland Barthes, *S/Z*, trans. Richard Miller (New York: Hill and Wang, 1974).
3. James Phelan's work, and more recently that of John Frow, can be seen as thoughtful attempts to bridge this divide—to see the interestingness of literary characters in the way they combine these two aspects of being rhetorical effects and constrained by their distinctive reference to real-world beings. This is what Uri Margolin means by the "ontological independence" of character—that characters on some level have a meaning that exceeds the language used to construct them. James Phelan, *Reading People, Reading Plots: Character, Progression, and the Interpretation of Narrative* (Chicago: University of Chicago Press, 1989); John Frow, *Character and Person* (Oxford: Oxford University Press, 2014); Uri Margolin, "Characterisation in Narrative: Some Theoretical Prolegomena," *Neophilologus* 67 (1983): 1–14.
4. See Alan Palmer, *Fictional Minds* (Lincoln: University of Nebraska Press, 2004); Lisa Zunshine, *Why We Read Fiction: Theory of Mind and the Novel* (Columbus: Ohio State University Press, 2006); and Blakey Vermeule, *Why*

Do We Care about Literary Characters? (Baltimore: Johns Hopkins University Press, 2011). For a critique of these positions, see the work of Omri Moses, for whom modernist characters are unique in the way they push against notions of "type" or "role" and instead privilege a sense of context and contingency. Omri Moses, *Out of Character: Modernism, Vitalism, Psychic Life* (Stanford, CA: Stanford University Press, 2014).

5. Deidre Lynch, *The Economy of Character: Novels, Market Culture, and the Business of Inner Meaning* (Chicago: University of Chicago Press, 1998); and David Brewer, *The Afterlife of Character, 1726–1825* (Philadelphia: University of Pennsylvania Press, 2011).

6. Character counts were derived using David Bamman's BookNLP tool using the NOVEL750 collection. See https://github.com/dbamman/book-nlp and script 5.1. Results are contained in the table Char_Counts_NovelEnglish700.csv.

7. Alex Woloch, *The One vs. the Many: Minor Characters and the Space of the Protagonist in the Novel* (Princeton, NJ: Princeton University Press, 2003).

8. Estimates on average number of characters are derived from the NOVEL750 data set. See script 5.1. This number rises slightly in the contemporary novel, to 89 (NOVEL_CONT). The estimate on the average number of novels in the period is based on Ted Underwood's work with the Hathi Trust data set: https://figshare.com/articles/Understanding_Genre_in_a_Collection_of_a_Million_Volumes_Interim_Report/1281251.

9. These averages are drawn from the NOVEL_CONT data set. See LIWC_Contemporary_Novels.csv.

10. For recent computational work on character, see David Bamman, Ted Underwood, and Noah Smith, "A Bayesian Mixed Effects Model of Literary Character," *Proceedings of the 52nd Annual Meeting of the Association for Computational Linguistics* (ACL, 2014): 370–379. They also have a forthcoming article on the declining coherence of gendered identity surrounding characters: Ted Underwood, David Bamman, and Sabrina Lee, "The Transformation of Gender in English-Language Fiction," *Cultural Analytics* (forthcoming). On the predictability of gender in the nineteenth century, see Matthew L. Jockers and Gabi Kirilloff, "Understanding Gender and Character Agency in the 19th Century Novel," *Cultural Analytics* (December 2016), article doi:10.22148/16.010, dataverse doi:10.7910/DVN/3UXBOJ.

11. Bamman, Underwood, and Smith, "A Bayesian Mixed Effects Model of Literary Character." For a system of characterization that looks at the perspectival construction of character—the way characterization is a product of narrative focalization—see the work of Fotis Jannidis, who describes a four-part system that includes reliability, mode, relevance, and directness. Each category offers one way of accounting for the conditional and perspectival nature of information surrounding characters. Fotis Jannidis, *Figur und Person: Beitrag zu einer historischen Narratologie* (Berlin: De Gruyter, 2004), 201.

12. Barthes, *S/Z*, 94; and Megan Ward, *Seeming Human: Artificial Intelligence and Victorian Realist Character* (Columbus: Ohio State University Press, 2018).
13. On "semic surplus," see Andrew J. Scheiber, "Sign, Seme, and the Psychological Character: Some Thoughts on Roland Barthes' *S/Z* and the Realistic Novel," *Journal of Narrative Technique* 21.3 (1991): 262–273, 265.
14. Jens Eder, Fotis Jannidis, and Ralf Schneider will speak of the "ontological incompletion of character" to emphasize the way characters' identity often resides beyond the words used to conjure them. See Jens Eder, Fotis Jannidis, and Ralf Schneider, eds., *Characters in Fictional Worlds: Understanding Imaginary Beings in Literature, Film, and Other Media* (Berlin: de Gruyter, 2010), 11. On the incompletion of character, see also Ruth Rosen, "Fictional Entities, Incomplete Beings," *Possible Worlds in Literary Theory* (Cambridge: Cambridge University Press, 1994), 108–143.
15. Barthes, *S/Z*, 92.
16. See in particular the work of Nancy Armstrong, *Desire and Domestic Fiction: A Political History of the Novel* (Oxford: Oxford University Press, 1990); Gillian Brown, *Domestic Individualism: Imagining Self in Nineteenth-Century America* (Berkeley: University of California Press, 1992); and Lynch, *The Economy of Character*.
17. All results in this section were generated by my student Hardik Vala, to whom I am very appreciative for his work on this project. We use a modified version of David Bamman's BookNLP. For a discussion, see Hardik Vala, David Jurgens, Andrew Piper, and Derek Ruths, "Mr. Bennet, His Coachman, and the Archbishop Walk into a Bar but Only One of Them Gets Recognized: On the Difficulty of Detecting Characters in Literary Texts," *Conference on Empirical Methods in Natural Language Processing* (EMNLP-2015).
18. We found that readers can identify aliases and resolve them to unique characters with upward of 98% accuracy. See Hardik Vala, Stefan Dimitrov, David Jurgens, Andrew Piper, and Derek Ruths, "Annotating Characters in Literary Corpora: A Scheme, the Charles Tool, and an Annotated Novel," *Proceedings of the Tenth International Conference on Language Resources and Evaluation* (LREC-2016). For estimates on computational co-reference resolution, see Bamman, Underwood, and Smith, "A Bayesian Mixed Effects Model of Literary Character."
19. We use the Stanford Dependencies parser, which can be accessed at http://nlp.stanford.edu/software/stanford-dependencies.shtml.
20. The other approach one can use is to attach all words within a given neighborhood of a target word (characters) and rank them according to some measure (for example, pointwise mutual information). This calculates each word's likelihood of appearing near the target word relative to its overall frequency in the text. "The" appears a great deal next to characters, but it also appears a great deal in texts more generally, so it does not score highly on such a test. The two disadvantages of this scheme are that a) infrequent words tend to be highly weighted; and b) it is very uncertain whether all

words in a neighborhood bear a meaningful relationship to the target word (where everything depends on how we define "meaningful").
21. You can try this yourself at http://nlp.stanford.edu:8080/corenlp/process.
22. According to the Stanford documentation, precision for the parser ranges from 69% for adjectival phrases to 99% for simpler constructions like possessives or subject/object-verb relations. We expect these numbers to be lower for more complex literary documents.
23. For this experiment, we take 1,000 random pairs of novels from each genre and calculate the KL-divergence between a) the character-text in each novel; and b) the character-text in the first novel and the non-character-text in the first novel. The top 1,000 words associated with either characters or the non-character space are selected with stopwords and communication verbs (*said, replied, answered,* etc.) removed. Novels are compared only to those within their own genre (prizewinners to prizewinners or nineteenth-century novels to nineteenth-century novels). We found that these results hold for all levels of vocabulary (from the top 100 words to the top 1,000). For the results, see KL_Test_WithinNovelsBetweenNovels_AllTrials.csv.
24. See KL_Test_WithinNovelsBetweenNovels_PerClass_Results.csv.
25. We limited vectors to the top 1,000 most frequent words associated with characters and nouns while controlling for stopwords and communication verbs. The similarity measure used was cosine similarity. We were able to replicate our results using total variational distance as well. For results, see Lexical_Diversity_Cosine_All.csv.
26. We test this using the top character from each novel and compare this to an averaged value for the top five nouns for the same novel. To calculate the similarity across narrative time, we calculate the cosine similarity between the character vector for the first half of the novel and the character vector of the same character for the second half of the novel, using the top 1,000 most frequent dependent words. We perform the same process for the top five nouns (i.e., compare each noun's first-half vector with its second) and take the average. These results were replicated when we examined only the top 100 most frequent words as well. See script 5.2. For results, see Lexical_Diversity_OverNarrativeTime.csv.
27. "Type" is a term that itself can be understood in numerous different ways. Here I use it to refer to a statistically meaningful categorization related to some character feature. For a typology of character types, that is, an overview of the different ways "type" has been understood with respect to character, see Jannidis, *Figur und Person*, 214.
28. The table containing all features for all novels is FeatureTable_All_Protagonist.csv.
29. See script 5.3a. The dictionaries can be found here: dict_communication.csv, dict_senseperception_stemmed.csv, dict_insight_stem.csv.
30. Results are contained in FeatureTable_ROM_VIC_Normalized.csv.
31. Overall, women (M = 13.4%, SD = 3.2%) tend to write more introverted char-

acters than men (M = 11.8%, SD = 3.2%), $t(274.04) = 4.623$, $p = 5.833\text{e-}06$. Both cogitative and perceptual lexica are significantly higher. All results are contained in FeatureTable_ROM_VIC_Labeled.csv. See script 5.3.

32. We can confirm these results using an analysis of variance (ANOVA) test, which indicates that gender has a significant effect on introversion: $F(3,415) = 9.49$, $p = 4.47\text{e-}06$. A Tukey's range test, which compares the means of each group to each other, indicates that the only group that has a significantly elevated level of introversion is the Female-Female group (female protagonists written by female authors). All other pairs indicate no significant difference. See script 5.3.
33. Armstrong, *Desire and Domestic Fiction*, 4.
34. Brown, *Domestic Individualism*.
35. Lynch, *The Economy of Character*, 151.
36. Hugh Stutfield, "The Psychology of Feminism," *Blackwood's Magazine* 161 (1897): 104.
37. For an elegant counter-reading of the value of female distance, see Patricia Frank, "The Ends of Aloneness: Scenes of Solitude in Nineteenth-Century Fiction," PhD diss., University of Wisconsin–Madison, 2015.
38. This is similar to the argument made by Matthew L. Jockers and Gabi Kirilloff, where the interiority of female characters is understood as a lack of access to physical action. See Jockers and Kirilloff, "Understanding Gender and Character Agency." For a strong counter-reading of this tradition that argues that domestic fiction actually engendered a new kind of "contractual subjectivity"—that is, the right of a woman to choose her spouse—see Wendy S. Jones, *Consensual Fictions: Women, Liberalism, and the English Novel* (Toronto: University of Toronto Press, 2005).
39. Brown, *Domestic Individualism*, 4–5.
40. As Lynch writes, "Since deep truths can so easily be represented as knowledge inaccessible to all but a few, the private reading could supply a means of restaging the stratification of society." Lynch, *The Economy of Character*, 146.
41. Catherine Gallagher, *Nobody's Story: The Vanishing Acts of Women Writers in the Marketplace, 1670–1920* (Berkeley: University of California Press, 1995), xxi.
42. See FeatureTable_ROM_VIC_Normalized_Female.csv.
43. The network was built in the following way: first, novels are reduced to only those sentences in which the main character appears. This allows us to avoid confusion, for example, if certain vocabulary is referring to a male character. Second, we identify collocates that appear within +/–9 words of one of the five selected keywords (which I represent using all possible cognates, so that "think" refers to the lexemes *thinking, thinks, think, thought,* and *thoughts*). Then, only those words are kept that are "distinctive" of female protagonists, meaning they are statistically more likely to appear in novels by women with female protagonists than in novels by men with

male protagonists. The terms are then weighted using a pointwise mutual information score, which calculates the probability of two words appearing together divided by their overall probability of appearing in the novel as a whole (and taking the log of that number). For example, *the* is very likely to appear near the word *look*, but because it appears frequently in the corpus as a whole, it receives a low PMI score. Because PMI favors low-occurring words, I use a threshold of 50 occurrences as the minimum number of times a word needs to appear in sentences in which female protagonists appear in novels by women. See script 5.4. Results are contained in Character_Collocates_Women_Distinctive_PMI.csv.

44. Results are based on a comparison of novels written by women with female protagonists and novels written by men with male protagonists. "Futurity" is measured through a custom dictionary that includes the words *will, forward, future, shall, ought, might,* and *must*. Odds ratios are calculated using a Fisher's exact test. Past and present tense are derived using the openNLP package in R. See script 5.5.
45. In order to arrive at these scores, I take only those sentences in which the female protagonist is present and the word *strong* appears. I then compare the word distributions in the sentences by George Eliot with Women Writers more generally using the Bing sentiment dictionaries. Results are located in MDW_Eliot_Strong.csv. See script 5.6.
46. Piper and Portelance, "How Cultural Capital Works."
47. For a full list, see MDW_Window.csv. Only words that appeared more than twenty times were kept, and a Fisher's exact test was used to calculate the increased odds of a word's occurrence in a sentence with a window.
48. Here emotion is calculated using the Bing Sentiment lexica. The odds ratio of positive words in sentences with the word window is 0.60 ($p < 2.2e-16$) and for negative words 0.79 ($p < 2.2e-16$), meaning we are 40% less likely to see positive words in the first case and about 21% in the second.
49. See Ferguson, "Now It's Personal"; and D. A. Miller, *Jane Austen: The Secret of Style* (Princeton, NJ: Princeton University Press, 2003), 92.
50. See script 5.8.
51. Underwood, Bamman, and Lee, "The Transformation of Gender."
52. The median amount of introversion for twentieth-century novels is 14.05%, while for nineteenth-century novels it is 12.96%, W = 4510000, $p < 2.2e-16$. A variance test indicates significantly higher levels of variance (2.1 times) in the twentieth-century collection, $F(3287) = 2.0604$, $p < 2.2e-16$. See script 5.9.
53. $F(5,1178) = 17.3$, $p < 2e-16$. Using a Tukey's range test, we see that Romance and YA are significantly lower than all other genres, while SciFi is significantly higher than Bestsellers (and YA and Romance), but interestingly not Prizewinners or Mysteries. See script 5.10.

CHAPTER SIX

1. "Corpus," *Lexikon des gesamten Buchwesens*, ed. Karl Löffler and Joachim Kirchner, vol. 1 (Leipzig, 1935–1937), 372.
2. "Procès-Verbal de l'ouverture du corps de J. J. Rousseau," *Oeuvres de J. J. Rousseau, avec des notes historiques*, vol. 22 (Paris: Lefèvre, 1819).
3. For an overview, see Matthew L. Jockers and Daniela M. Witten, "A Comparative Study of Machine Learning Methods for Authorship Attribution," *Literary and Linguistic Computing* 25.2 (2010): 215–224; and Efstathios Stamatatos, "A Survey of Modern Authorship Attribution Methods," *Journal of the American Society for Information Science and Technology* 60.3 (March 2009): 538–556.
4. While initial work emphasized the importance of so-called stopwords, more recent work has suggested that extending the range of words can improve attributional performance. See David Hoover, "Testing Burrows' Delta," *Literary and Linguistic Computing* 19.4 (2004): 453–475; and David Hoover, "Delta Prime?," *Literary and Linguistic Computing* 19.4 (2004): 477–495.
5. See David Hoover, "Corpus Stylistics, Stylometry, and the Styles of Henry James," *Style* 41.2 (2007): 174–203.
6. Hierarchical clustering is a method where similarities between documents are generated in a hierarchical, treelike way. It begins by separating the documents into two clusters, then three, then four, and so on, until it arrives at the simplest level, where each cluster represents a pair—a document and its nearest neighbor. Following the tree upward or downward allows us to cluster documents into more or less general combinations. The traditional way of representing this process is through the use of a dendrogram. A fuller discussion of the model will come later in the chapter and is fully described in appendix B. For the implementation of the model on Whitman, see script 6.1.
7. This is a prime example of the challenges of "validation" in computational literary studies. Is this model highlighting something distinctly new in the scholarship—and contradicting scholarly consensus in the process that is in some sense "flawed"? Or does the contradiction highlight that the model itself is flawed if it cannot agree with generations of Whitman scholars? I think the key point here, to reiterate my methodological refrain from the introduction, is the way the value of the model lies in the way it makes explicit the conditions of its judgments, whereas traditional criticism tends to operate with more black boxes. There is no absolutely right or wrong answer to the question of Whitman's periodization, only different ways of assessing it.
8. A "page" consists of 300 words, which approximates the average page length of both the first and final editions of *Leaves of Grass*. Because the actual page lengths vary considerably across editions, it is important to use a standardized page for the model. I will discuss the way "similarity" is

measured in greater detail in the following section, but it consists of four main dimensions of lexical, semantic, phonetic, and syntactic features. See script 6.2 for implementation.

9. Because the length of Whitman's editions increases considerably over the course of his career, I tested (b) and (c) while controlling for the number of pages backward a page can be most similar to (limited to most recent 100 pages). It's possible that as the editions lengthen, there is an increased likelihood that a page could "befriend" a page that is less than five pages away or more than twenty. However, doing so results in no overall change in the larger trend shown here.

10. "Vulnerability" is a traditional measure used in network science to assess how well connected a graph is (i.e., how vulnerable it is to attack or breakdown). It consists of removing nodes through a process known as percolation, usually in descending order of connectivity (so the most connected node is removed first, etc.), until the graph is not fully connected (meaning at least one node is no longer connected to any other node). In my model, I use a breakpoint when there are two components remaining where the smaller is at least 50% as large as the larger component. The idea is to measure how long it takes before the poet's corpus divides itself into two relatively coherent entities, that is, until s/he becomes double. For an introduction, see M. E. J. Newman, "Percolation and Network Resilience," in *Networks: An Introduction* (Oxford: Oxford University Press, 2010), 592–625.

11. Thomas K. Landauer, Peter W. Foltz, and Darrell Laham, "An Introduction to Latent Semantic Analysis," *Discourse Processes* 25.2–3 (1998): 259–284.

12. You can use script 6.3 to test out the results of the different models discussed here using other poets.

13. For an introduction, see Newman, "Percolation and Network Resilience."

14. For a full implementation, see script 6.4. Results are contained in Vulnerability_Global_All.csv.

15. For a full implementation, see script 6.5. Results are contained in Vulnerability_Local_All.csv.

16. To test the significance of a poem's vulnerability, I use what is called a randomization test. Instead of keeping the poems in order, I randomly shuffle this order so that the temporal structure of the poet's career no longer exists. I then calculate the average similarity of each poem in this new fictional model to all the poems that precede it, just as in the regular model. I then perform this operation of randomly shuffling the poems' order 200 times and calculate the average similarity of each fictional poem's similarity to those prior to it for every poem. The dotted line represents the values that appear in only 1% of all the random trials for each poem, meaning that if a poem is above or below that amount, it happened in less than 1% of all random permutations of the data, that is, is exceedingly rare and thus not accountable for by random fluctuation.

17. Foote, "Automatic Audio Segmentation." For an application of this tech-

nique to understanding changes in musical style in popular music, see Mauch et al., "The Evolution of Popular Music." For a response, see Underwood et al., "You Say You Found a Revolution."

18. In order to calculate Foote novelty, I use the following steps. First, a poet's corpus is reduced to the feature table described above. Then a similarity table between each poem and every other poem is generated using cosine similarity. Foote novelty then uses a sliding transformation across the data that magnifies differences before and after a given target poem. So, as it moves from poem 1 to 2 to 3, it transforms the similarity scores of a certain number of poems before and after the target poem to accentuate the differences between them. (Imagine sliding a smaller matrix down the diagonal of a similarity matrix.) As the novelty score rises, we are seeing an increase in difference between poems that occur after the midpoint of the window and those that occur before. A rise in the curve is thus designed to capture not just a single anomalous divergence, but a sense of movement away from a previous stylistic norm. The outcomes of the Foote measures have been smoothed to avoid more local changes, given that we want a period to exceed the window being observed. This helps reduce noise of smaller more minor periods of variation. For smoothing, I have used a rolling mean on a window of 1/10 the number of poems in the collection ($k = n/10$, where n = number of poems). In order to calculate the significance threshold, I once again permute the poems 200 times and run the same process. I use the 95th percentile value, meaning that only those values that exceed 5% of all cases in the random permutations of the data are considered "significant," that is, warrant being called a "period." Whereas Underwood et al. recommend permuting the data along the diagonal axes of the matrix for historical data (in order to approximate the high degree of historical continuity found in yearly stylistic data), I find that poets' careers are sufficiently nonlinear not to require this. Permuting the entire data set randomly while maintaining the underlying symmetrical structure (of the similarity matrix) should suffice to capture when a poet's stylistic variability achieves a significant degree of change. See Underwood et al., "You Say You Found a Revolution." See script 6.6.

19. For a list of statistics relative to periods, see FooteNovelty_Results_All.csv. See the table FooteNovelty_Poems_All.csv for lists of poems that mark period shifts in each poet's career. For the calculations discussed in the text, see script 6.7.

20. Edward Said, *On Late Style: Music and Literature Against the Grain* (New York: Vintage, 2007).

21. For a history of other work on late style as well as a critique of Said's formulation, see Linda Hutcheon and Michael Hutcheon, "Late Style(s): The Ageism of the Singular," *Occasion: Interdisciplinary Studies in the Humanities*, vol. 4 (May 31, 2012), http://arcade.stanford.edu/occasion/late-styles-ageism-singular.

22. In order to test for significance, I use a randomization test for each measure for every poet. What this means is that I first calculate the measure given the actual division of the corpus into late and not-late periods. I then randomly permute the linear order of the corpus 1,000 times and calculate the measure on each of these fictional careers. A change is deemed significant if it happens in less than 5% of all random cases; that is, the difference we are seeing between the values in the late versus earlier periods is highly unlikely to be the result of chance.
23. Tuldava's measure is a modification of Flesch Reading Ease, which is widely used in selecting texts for standardized testing. Like most readability metrics, it combines measures of sentence length with word length (using syllables, not number of characters). The exact formula is (Syllables/Words) * logn(Words/Sentences). Because it uses a logarithmic scale, it is less susceptible to the huge swings caused by outliers (texts with little punctuation, for example, which are very common in poetry). This also makes it more appropriate to compare poems from different languages. See Peter Grzybek, "Text Difficulty and the Arens-Altmann Law," *Text and Language: Structures—Functions—Interrelations: Quantitative Perspectives*, ed. Peter Grzybek, Emmerich Kelih, and Ján Mačutek (Vienna: Praesens, 2010), 57–70. See script 6.8 and Difficulty_Results_All.csv.
24. I measure syntactic irregularity using a summed tf-idf score for 1–3gram parts of speech for each poem. For the entire corpus, each part-of-speech gram up to 3 in a row is weighted according to its in-document frequency and between-document rarity. These scores are then added together for each poem. The more often a poem uses a syntactic pattern that occurs less often in the corpus, the higher the overall tf-idf score will be for a given poem. I then compare the distributions of these scores for the late period compared to the rest of the career. See script 6.9 and Syntactic_Irregularity_All.csv.
25. I use a modified version of type-token ratio to capture vocabulary richness. In order to control for differing poem lengths, for each period (late, not late) I take 100 samples of 500 contiguous word units and calculate the median type-token ratio. These actual values are then compared to the randomly permuted values to see if there is a significant increase or decline in vocabulary richness in the late period. See script 6.10 and TTR_All_Late.csv.
26. Lancashire and Hirst, "Vocabulary Changes in Agatha Christie's Mysteries."
27. Concreteness is defined as the ratio of physical entities to abstractions for each sample (late, not late). Poems are first reduced only to their nouns using a part-of-speech tagger in R, and then the nouns are subsequently transformed into their hypernym trees using WordNet. Every noun will thus contain at least one count of the final entity in the tree (physical entity or abstraction). Python scripts to run these transformations in French, German, and English are located in Hypernym_Translator. See script 6.12 and Concreteness_All.csv.
28. Generality is measured as the percentage of words in a given group that

are hypernyms of the words in the other group. First, poems are rendered as their hypernym trees using the scripts in the note above. Then both the lexical and hypernym representations of a poet's corpus are loaded and compared to each other. First, I calculate what percentage of words in the late period are also in the hypernyms of the early period (meaning these words are hypernyms of the early period words). I then calculate what percentage of words in the early period are in the hypernyms of the late period. This gives me a hypernymy score for each period. I then subtract the early value from the late value to get an overall late-period increased hypernymy score. This is then compared to the permuted values from the randomization test to see if this level is indeed significant. See script 6.13 and Generality_All.csv.
29. Hutcheon and Hutcheon. "Late Style(s)."
30. Biplots are useful to observe the relationships between variables and observations, or in our case poets' careers and measures of late style. The positions of the poets are based on the first two principal components of the late style features after the features have been normalized (turned into z-scores). The vectors (arrows) represent the locations of the features relative to the poets' careers. The closer two vectors are, the more they are correlated. In this graph, we see how each feature points in a different direction, suggesting the low levels of correlation between the variables. See script 6.14 for more details.
31. Counts derived from the MLA Bibliography as of December 16, 2016.
32. Vocabulary diversity is measured using type-token ratio across 1,000 random samples of 500-word windows drawn from a poet's entire career. See script 6.11 and TTR_All.csv.
33. Tyler T. Schmidt, "'Womanish' and 'Wily': The Poetry of Wanda Coleman," *Obsidian III: Literature in the African Diaspora* 6.1 (2005): 134.
34. For a reading of Coleman's late work, see Jennifer Ryan-Bryant, "Saying Goodbye: Elegiac Subjectivity in Wanda Coleman's 'The World Falls Away,'" *Hecate: A Women's Interdisciplinary Journal* 40.1 (2014): 116–131.
35. Sentences are derived by using the sentence annotator in openNLP in R. A mean sentence length is then calculated for every poem, with the median values for each period being compared. The median per-poem average sentence length is 37 in the not-late work and 29 in the late work. If we randomly sample from her non-vulnerable poems the same amount of poems as in the vulnerable set, less than 1% of the time will we find a median sentence length that is lower than that of the vulnerable poems. See script 6.15.
36. Punctuation is calculated for only the following marks: question, exclamation, period, comma, semicolon, colon, and parentheses. The median rate of punctuation in the vulnerable poems is about 7.9 marks for every 100 words, while for the non-vulnerable group it is closer to 4.6. In 1,000 random samples, we never find a group of non-vulnerable poems whose median value approaches this rate. See script 6.16.

37. Vocabulary innovation is calculated as the percentage of word types in the vulnerable poems that are not present in the rest. See script 6.17.
38. Parts of speech are captured through the openNLP annotator in R. Distinctive parts of speech are identified using a Fisher's exact test. See script 6.18.
39. Concreteness and generality are calculated as above. See scripts 6.19 and 6.20. Her language is considerably more concrete in the vulnerable poems, with a 58% increase in the ratio of physical to abstract terms. If we run a permutation test and randomly shuffle the corpus 1,000 times, once again we never see this degree of change in terms of the concretization of language. No significant difference is shown with respect to her level of generality or specificity in the vulnerable poems.
40. "There is" is 47% more likely to appear in vulnerable poems than non-vulnerable ones ($p = 0.0086$). See script 6.18.
41. As Levinas writes, "L'indétermination en fait l'actuité. Il n'y a pas d'être déterminé, n'importe quoi vaut pour n'importe quoi. Dans cette équivoque se profile la menace de la présence pure et simple, de l'il y a." Emmanuel Levinas, *De l'existence à l'existant* (Paris: Librairie Philosophique, 2002), 96.
42. Rebecca Couch Steffy, "Consuming Figures: Wanda Coleman, Writing, and the Flesh," *Hecate: A Women's Interdisciplinary Journal* 40.1 (2014): 80–96.

CONCLUSION

1. The quotations are drawn from the following sources: Timothy Snyder, *Bloodlands: Europe Between Hitler and Stalin* (New York: Basic Books, 2010); Gary Krist, *Empire of Sin: A Story of Sex, Jazz, Murder and the Battle for Modern New Orleans* (New York: Broadway Books, 2015); and Doris Kearns Goodwin, *The Bully Pulpit. Theodore Roosevelt, William Howard Taft, and the Golden Age of Journalism* (New York: Simon and Schuster, 2013).
2. This model is based on a sample of 200 works of bestselling contemporary histories according to Amazon.com sales rankings. The model was built using Ben Schmidt's word2vec package in R. See script C.0.
3. The citations are from Francis Fawkes, "A Description of May" (1720); Richard Glover, "Leonidas" (1712); and Henry Brooke, *Universal Beauty: A Poem* (1735).
4. This citation is from William Camden's *Britain, or, A chorographicall description of England, Scotland, and Ireland*, trans. Philemon Holland (1610), and marks the first citation listed in the *OED* for the verb "to implicate."
5. Marjorie Levinson, "The New Historicism," in *Rethinking Historicism: Critical Readings in Romantic History* (Hoboken, NJ: Blackwell, 1989), 20, 34.
6. See script C.1 for further description.
7. The sample consists of 2,745 articles acquired through the JSTOR Data for Research platform. It contains all "research articles" published between 2005 and 2014 listed within the discipline of "language and literature," on

the subject of "language and literature," in English, and between 8 and 30 pages in length.
8. Distinctiveness is calculated using log-likelihood ratios. See script C.3.
9. For trial results, see script C.4. This process essentially builds a vector space model of all the documents in the collection and identifies those articles most similar to my chapters. One could experiment with dimension-reduction techniques such as LDA or LSA to rank associations that more nearly match scholarly expectations.
10. As James A. Evans and Jacob G. Foster write, "The ecology of modern scientific knowledge constitutes a complex system: apparently complicated, involving strong interaction between components, and predisposed to unexpected collective outcomes." For an introduction to the field, see James A. Evans and Jacob G. Foster, "Metaknowledge," *Science* 331 (February 2011): 721–725, 724.
11. Andrew Goldstone and Ted Underwood, "The Quiet Transformations of Literary Studies: What Thirteen Thousand Scholars Could Tell Us," *New Literary History* 45.3 (2014): 359–384.
12. Mark Algee-Hewitt, "The Shape of Reading" (forthcoming).
13. This data represents 1,937 and 6,252 bibliographic records in the field of literary studies from the years 1970 and 2015, respectively. The data was downloaded from the MLA database using the ProQuest interface in January 2017. When asked, the MLA would not share this data from their own records. See script C.5.
14. An alternative way to measure this, which would take into account the value of that expanded number of authors, would be to use the Herfindahl-Hirschmann Index (HHI). This is used in economics to measure the concentratedness of market share within an industry. The more actors, the less concentrated an industry is assumed to be. Here we see how the HHI scores have decreased considerably since 1970, from 0.0095 to 0.0024.
15. Chad Wellmon and Andrew Piper, "Publication, Power, and Patronage: On Inequality and Academic Publishing," *Critical Inquiry* (Summer 2017), http://criticalinquiry.uchicago.edu/publication_power_and_patronage_on_inequality_and_academic_publishing/.
16. Data and code can be accessed at https://doi.org/10.6084/m9.figshare.4558072.v3.
17. The top 20% of institutions still account for over 80% of all articles, while the top ten institutions now account for only 29.5% of articles (compared to just over 50% for PhDs).

Index

This index was generated using the script indexer.R found in the code repository.

Amanda und Eduard. See Mereau, Sophie
Anton Reiser. See Moritz, Karl Philip
Aragon, Louis, 162–63, 170
Aristotle, 68–70, 94
Arnim, Bettina von, 78
Audet, Martine, 169, 170
Auerbach, Erich, 6–12
Augustine of Hippo, 1, 3, 20, 45–48, 60
Austen, Jane, 43, 45, 58, 62, 122–24, 126, 131, 133, 135, 136, 143, 144
authorship, 6, 147–48

Baraka, Amiri, 23, 30, 36–39, 153
Barthes, Roland, 1, 13, 16, 121, 122
Batailles, Georges, 23, 24, 27–29. *See also* excess
beauty, 37, 80–92
bestsellers, 126, 127, 129, 145
Brentano, Clemens, 78, 81
Brossard, Nicole, 153
Burney, Frances, 142

Campe, Joachim, 45, 58, 60, 192
canon, the, 1, 23, 44, 58, 98, 99, 100, 102, 107, 108, 133, 140
Castle, The. See Kafka, Franz
Celan, Paul, 153, 170
Cervantes, Miguel de, 44, 57, 62
classification, 100–105, 127

close reading, 10, 83, 143, 144. *See also* too close reading
clustering, 47, 85, 87, 163, 165
Coleman, Wanda, 153, 172–77
commonplacing, 3, 68–75
Confessions. See Augustine of Hippo
configuration, 15, 19, 67, 68, 81, 83, 85, 87, 93, 148
connotation, 53, 56. *See also* synonymy
corporality, 36–38, 74, 82, 100, 110, 111, 114, 116–19, 121, 131, 145–53, 167–73. *See also* sense perception
cosine similarity, 14, 15, 17, 199–201; validation of, 154–58
Curtius, Ernst Robert, 66, 67, 92
Cusk, Rachel, 118, 130, 131

death, 31, 35, 60, 68–70, 75–91. *See also* life
Deleuze, Gilles, 2, 12, 13, 75
dependency parser, 125
Derrida, Jaques, 97
diagrams, theory of, 18–21
difficulty: in poetry, 167–72
distinctive words, 34, 59, 81
distributional semantics, 13–18, 43, 48, 66
Don Quixote. See Cervantes, Miguel de
Droste-Hülshoff, Annette von, 159–65, 171

INDEX

Eichendorff, Joseph von, 171
Eliot, George, 48, 118, 135, 139, 140, 148, 183
Emma. See Austen, Jane
encounter, in the novel, 90–92
epistolary novel, the, 69, 75–93
Erasmus of Rotterdam, 71–73
excess: in poetry, 29–41; theory of, 27–29
exile, 6–8, 11, 19, 166–68
extroversion: in novels, 130–43. See also introversion

figuration. See configuration
Fontane, Theodor, 48–53, 56, 58, 60–62, 192
foote novelty, 62–63, 164–67
Foucault, Michel, 68, 75, 85, 92
Frapan, Ilse, 78, 82
From the Earth to the Moon. See Verne, Jules

Gaskell, Elizabeth, 141
Genette, Gérard, 13, 43, 64
genre, and the novel, 43–45; and the novel of lack, 45–61
Goethe, J. W.: and the novel of lack, 45, 58, 61, 62, 72; and poetic periods, 164–66; and poetry, 148, 153, 171, 172; and the theme of life and death, 78–81

Heaney, Seamus, 153
Hemans, Felicia, 162–63
heteroglossia, 45, 50
histogram, definition of, 29–30
Hölderlin, Friedrich, 78, 80–82
Hugo, Victor, 161, 166–67
humanism, 4–5
Hyperion. See Hölderlin, Friedrich
hypernyms, 32–36, 170
hypotheses, 11–14, 28, 58, 80, 92, 99, 100, 110, 115, 116

interiority. See introversion
introversion: in novels, 130–46
Irrungen, Wirrungen. See Fontane, Theodor

Jackson, Angela, 35
James, Henry, 62–63
Jean Paul, 82
Joyce, James, 1, 20, 26, 48; and the novel of lack, 53–56, 60–61

Kafka, Franz, 2, 16–18, 50–52, 64
Kirsch, Sarah, 169

Lasker-Schüler, Else, 153, 162, 170
late style, 167–72. See also difficulty
Latour, Bruno, 12, 76, 92
Lesabéndio. See Scheerbart, Paul
life, 68–70, 75–91. See also death
liquidity, in the novel, 79–83

machine learning, 2, 5, 6, 19, 97, 101, 104, 117, 126, 127
Malte Laurids Brigge. See Rilke, Rainer Maria
McClane, Kenneth, 41
Mereau, Sophie, 80, 87–93
Mill on the Floss. See Eliot, George
modeling: theory of, 6–12, 42–45, 66–70; and visualization, 18–21. See also topic modeling; vector space models
modernism, 20, 53, 56, 78, 99, 166
Morgan, Edwin, 34–36
Moritz, Karl Philip, 48–49
Morrissey, Judd, 1–3
Mrs. Dalloway. See Woolf, Virginia
Munro, Alice, 93–96

narrative discourse, 43, 61, 62
negation, 34, 60, 80, 100, 113, 114, 165
Noël, Marie, 171
nostalgia, 139, 140, 143
novel, the: of encounter, 90–93; the great reversal in, 115–17; of lack, 45–61; of life and death, 75–92; and phenomenology, 105–17. See also bestsellers; epistolary novel, the; prizewinning novels

Outline, the novel. See Cusk, Rachel

phenomenology. See sense perception
phonemes, 56, 148, 154
Plath, Sylvia, 170–71
polysemy, 50, 56, 61
Ponge, Francis, 153
Portrait of a Lad. See James, Henry
poststructuralism, 13, 92, 97, 100
Pride and Prejudice. See Austen, Jane
prizewinning novels, 126, 127, 129, 145
pronouns, 33, 42, 95, 96, 102, 106–11, 113, 120, 124
Pynchon, Thomas, 119, 124

Radcliffe, Ann, 122, 136, 138, 139, 140
realism, 27, 50, 99
repetition, 1–6, 18, 35, 43, 56, 93, 110, 157

242

representation, 6–12. *See also* modeling, theory of
rereading, 1–3, 12
resolution, 6, 12, 26, 43, 141–44
Rich, Adrienne, 153, 161
Rilke, Rainer Maria, 58, 78, 82–83
Robinson, Mary, 136, 171
Robinson der Jüngere. See Campe, Joachim
Romanticism, 91, 139, 142
Rossetti, Christina, 153, 171
Rukeyser, Muriel, 154–58, 166, 170, 171

Scheerbart, Paul, 23–28, 48, 57
Schiller, Friedrich, 171
science fiction, 19, 23, 57, 113, 127, 145, 146
Searle, John, 94–97
sense perception: in the novel, 10, 19, 99, 100, 105–17
social network analysis, 49–53, 62, 83–85, 136–38, 145–46, 149–53, 159–62
Sorrows of Young Werther, The, 14–15, 26, 78–82
Stein, Gertrude, 20
Stoddard, Elizabeth, 135

Tieck, Ludwig, 78, 80
too close reading, 143
topic modeling, 66–70
topics, history of, 70–75
topological reading, 66–70, 75–93

Ulysses, 1, 26, 48, 53–56, 192

vector space models, 6–9, 14–17, 39–40, 46, 48, 62, 85, 104, 125, 129, 154
Verhaeren, Émile, 166–67
Verne, Jules, 45, 48, 57, 59
visuality: of academic scholarship, 178–85; and diagrams, 18–21; and reading, 66–70; and women in novels, 138–43

Whitman, Walt, 37, 42, 148–53, 159
Wilhelm Meister's Travels, 45, 61
Williams, William Carlos, 28, 171
windows, 138–43
Woolf, Virginia, 45, 58–61, 118, 125–26
WordNet, 33, 86, 170
Work. See Frappan, Ilse

Young, Al, 163